Current Developments in Mathematics 2009

Edited by

David Jerison
Barry Mazur
Tomasz Mrowka
Wilfried Schmid
Richard P. Stanley
Shing-Tung Yau

International Press
www.intlpress.com

Current Developments in Mathematics, 2009

2010 Mathematics Subject Classification. 00Bxx

ISBN 978-1-57146-347-0

Paperback reissue 2017. Originally published in 2010 under ISBN 1-57146-146-9 (casebound).

Printed in the United States of America.

Contents

Survey on the Fundamental Lemma
Ngô Bao Châu 1

The Arf-Kervaire Invariant Problem in Algebraic Topology: Introduction
Michael A. Hill, Michael J. Hopkins, and Douglas C. Ravenel 23

Phase Transitions, Minimal Surfaces and A Conjecture of De Giorgi
O. Savin 59

Recent Developments in Mathematical Quantum Chaos
Steve Zelditch 115

Current Developments in Mathematics, 2009

Survey on the Fundamental Lemma

Ngô Bao Châu

This is a survey on the recent proof of the fundamental lemma. The fundamental lemma and the related transfer conjecture were formulated by R. Langlands in the context of endoscopy theory for automorphic representations in [**26**]. Important arithmetic applications follow from the endoscopy theory, including the transfer of automorphic representations from classical groups to linear groups and the construction of Galois representations attached to automorphic forms via Shimura varieties. Independent of applications, endoscopy theory is instrumental in building a stable trace formula that seems necessary to any decisive progress toward Langlands' conjecture on functoriality of automorphic representations.

There are already several expository texts on endoscopy theory and in particular on the fundamental lemma. The original text [**26**] and articles of Kottwitz [**19**], [**20**] are always the best places to learn the theory. The two introductory articles to endoscopy, one by Labesse [**24**], the other [**14**] written by Harris for the Book project are highly recommended. So are the reports on the proof of the fundamental lemma in the unitary case written by Dat for Bourbaki [**7**] and in general written by Dat and Ngo Dac for the Book project [**8**]. I have also written three expository notes on Hitchin fibration and the fundamental lemma : [**34**] reports on endoscopic structure of the cohomology of the Hitchin fibration, [**36**] is a more gentle introduction to the fundamental lemma, and [**37**] reports on the support theorem, a key point in the proof of the fundamental lemma written for the Book project. The survey follows the same plan as [**36**] but more details have been added.

This report is written when its author enjoyed the hospitality of the Institute for Advanced Study in Princeton. He acknowledges the support of the Simonyi foundation and the Monell Foundation during his stay in the Institute.

1. Orbital integrals over non-archimedean local fields

1.1. First example. Let V be a n-dimensional vector space over a non-archimedean local field F, for instant the field of p-adic numbers. Let $\gamma : V \to V$ be a linear endomorphism with distinct eigenvalues in an algebraic closure of F. The centralizer I_γ of γ is of the form

$$I_\gamma = E_1^\times \times \cdots \times E_r^\times$$

where $E_1, \ldots, E_r$ are finite separable extensions of F. This is a commutative locally compact topological group.

Let $\mathcal{O}_F$ denote the ring of integers in F. We consider the set of lattices of V that are sub-$\mathcal{O}_F$-modules $\mathcal{V} \subset V$ of finite type with maximal rank. We are interested in the subset $\mathcal{M}_\gamma$ of lattices $\mathcal{V}$ of V such that $\gamma(\mathcal{V}) \subset \mathcal{V}$. The group I_γ acts the set $\mathcal{M}_\gamma$. This set is infinite in general but the set of orbits under the action of I_γ is finite. We fix a Haar measure dt on the locally compact group I_γ. We consider a set of representatives of orbits of I_γ on $\mathcal{M}_\gamma$ and for each x in this set, let denote $I_{\gamma,x}$ the compact open subgroup of I_γ of elements stabilizing x. The finite sum

$$\sum_{x \in \mathcal{M}_\gamma / I_\gamma} \frac{1}{\mathrm{vol}(I_{\gamma,x}, dt)} \tag{1}$$

is a typical example of orbital integrals.

1.2. Another example. A basic problem in arithmetic geometry is the determination of the number of abelian varieties equipped with a principal polarization defined over a finite field $\mathbb{F}_q$. The set of isogeny classes of abelian varieties over finite fields is described by Honda and Tate. As usual, we first describe the set the principally polarized abelian varieties that are equipped with an isogeny to a fixed one in requiring that the isogeny be compatible with the polarization. We will be concerned only with ℓ-polarization for some fixed prime number ℓ which is different from the characteristic of $\mathbb{F}_q$.

Let A be a n-dimensional abelian variety over a finite field $\mathbb{F}_p$ equipped with a principal polarization. The $\mathbb{Q}_\ell$-Tate module of A

$$T_{\mathbb{Q}_\ell}(A) = \mathrm{H}_1(A \otimes \bar{\mathbb{F}}_p, \mathbb{Q}_\ell)$$

is a $2n$-dimensional $\mathbb{Q}_\ell$-vector space equipped with

- a non-degenerate alternating form that is induced by the polarization,
- a Frobenius operator σ_p induced from the $\mathbb{F}_p$-structure of A,
- a self-dual lattice $T_{\mathbb{Z}_\ell}(A) = \mathrm{H}_1(A \otimes \bar{\mathbb{F}}_p, \mathbb{Z}_\ell)$ which is stable under σ_p.

Let A' be a principally polarized abelian variety equipped with a ℓ-isogeny to A compatible with polarizations and defined over k. This isogeny induces an isomorphism between $\mathbb{Q}_\ell$-vector spaces $T_{\mathbb{Q}_\ell}(A)$ and $T_{\mathbb{Q}_\ell}(A')$ that is compatible with symplectic forms and Frobenius operators. Defining this ℓ-isogeny is thus equivalent to defining a self-dual lattice $\mathrm{H}_1(A', \mathbb{Z}_\ell)$ of $\mathrm{H}_1(A, \mathbb{Q}_\ell)$ stable under σ_p. Orbital integral for symplectic group enters in this way in the solution of the problem of counting the number of principally polarized abelian varieties over finite field within a fixed isogeny class.

The description of the set of p-isogenies where p is the characteristic of $\mathbb{F}_q$ is more complicated. The solution is based on the crystalline cohomology of the abelian variety instead of etale ℓ-adic cohomology, and can be translated into semi-linear algebra instead of linear algebra. Instead of orbital integral, the answer is expressed naturally in terms of twisted orbital integrals. Moreover, the test function is no longer the unit of the Hecke algebra but the characteristic function of the double class indexed a the minuscule coweight of the group of symplectic similitudes.

As isogeny is required to be compatible with polarization, the classification of principally polarized abelian varieties can't be immediately reduced to the classification of Honda and Tate. There is indeed a subtle difference between requiring A and A' to be isogenous or A and A' equipped with polarization to be isogenous. In [**23**], Kottwitz observed that this difficulty is of endoscopic nature. He expressed the number of points with values in a finite field on Siegel's moduli space of polarized abelian varieties in terms of orbital integral and twisted orbital integrals in taking into account the endoscopic phenomenon. He proved in fact the same result for a larger class of Shimura varieties classifying abelian varieties with polarization, endomorphisms and level structures.

1.3. General orbital integrals. Let G be a reductive group over F, $\mathfrak{g}$ its Lie algebra. Let γ be an element of $G(F)$ or $\mathfrak{g}(F)$ which is strongly regular semisimple in the sense that its centralizer I_γ if a F-torus. Choose a Haar measure dg on $G(F)$ and a Haar measure dt on $I_\gamma(F)$.

For $\gamma \in G(F)$ and for any compactly supported and locally constant function $f \in C_c^\infty(G(F))$, we set

$$\mathbf{O}_\gamma(f, dg/dt) = \int_{I_\gamma(F)\backslash G(F)} f(g^{-1}\gamma g)\frac{dg}{dt}.$$

We have the same formula for an element of the Lie algebra $\gamma \in \mathfrak{g}(F)$ and for $f \in C_c^\infty(\mathfrak{g}(F))$. By definition, the orbital integral $\mathbf{O}_\gamma$ does not depend on γ but only on its conjugacy class. It also depends on the choice of Haar measures dg and dt.

We are mostly interested in the unramified case. We assume that G has a reductive model over $\mathcal{O}_F$ i.e. there exists a reductive group scheme over $\mathcal{O}_F$ whose generic fiber is G. This is the case for instant for Chevalley groups. We use a slight abuse of notation in assigning also the letter G to the reductive group scheme over $\mathcal{O}_F$. The group $K = G(\mathcal{O}_F)$ of integral points is a maximal compact subgroup of $G(F)$. We choose the Haar measure dg such that K has volume one. Consider the set

$$\mathcal{M}_\gamma = \{x \in G(F)/K \mid gx = x\}, \tag{2}$$

equipped with an action of $I_\gamma(F)$. The orbital integral of the characteristic function 1_K of K admits a concrete description

$$\mathbf{O}_\gamma(1_K, dg/dt) = \sum_{x\in I_\gamma(F)\backslash \mathcal{M}_\gamma} \frac{1}{\mathrm{vol}(I_\gamma(F)_x, dt)} \tag{3}$$

where x runs over a set of representatives of orbits of $I_\gamma(F)$ in $\mathcal{M}_\gamma$ and $I_\gamma(F)_x$ is the compact open subgroup of $I_\gamma(F)$ stabilizer of x. The function 1_K also called the unit of the Hecke algebra plays a very special role in the global setting.

If $G = \mathrm{GL}(n)$, the space of cosets $G(F)/K$ can be identified with the set of lattices in F^n so that we recover the lattice counting problem in the first example. For classical groups, orbital integrals for the unit function can also be expressed as the number of selfdual lattices fixed by an automorphism.

1.4. The Arthur-Selberg trace formula. We consider now a semisimple group G defined over a global field F that can be either a number field or the field of rational functions on a curve defined over a finite field. It is of interest to understand the trace of Hecke operator on automorphic representations of G. The Arthur-Selberg trace formula is a powerful tool for this quest. It has the following form

$$\sum_{\gamma\in G(F)/\sim} \mathbf{O}_\gamma(f) + \cdots = \sum_\pi \mathrm{tr}_\pi(f) + \cdots \tag{4}$$

where γ runs over the set of anisotropic conjugacy classes of $G(F)$ and π over the set of discrete automorphic representations. The trace formula contains also more complicated terms related to hyperbolic conjugacy classes on one side and the continuous spectrum on the other side.

The test functions f are of the form $f = \otimes f_v$ with f_v being the unit function in Hecke algebra of $G(F_v)$ for almost all finite places v of F. The global orbital integral

$$\mathbf{O}_\gamma(f) = \int_{I_\gamma(F)\backslash G(\mathbb{A})} f(g^{-1}\gamma g)dg$$

is convergent for isotropic conjugacy classes $\gamma \in G(F)/\sim$. After choosing a Haar measure $dt = \bigotimes dt_v$ on $I_\gamma(\mathbb{A})$, we can express the above global integral as follows

$$\mathbf{O}_\gamma(f) = \mathrm{vol}(I_\gamma(F)\backslash I_\gamma(\mathbb{A}), dt) \prod_v \mathbf{O}_\gamma(f_v, dg_v/dt_v).$$

Local orbital integral of semisimple elements are convergent for every v and are equal to one for almost all v if the measure dt is chosen so that $I_\gamma(\mathcal{O}_v)$ has volume one for almost all v. The torus I_γ has an integral form that is well defined up to finitely many places. The volume term is finite when the global class γ is anisotropic.

Arthur introduced truncation operator to deal with the continuous spectrum and with non isotropic conjugacy classes. In his geometric expansion, Arthur has more complicated local integral that he calls weighted orbital integrals, see [**1**].

1.5. Shimura varieties. Similar strategy has been used for the calculation of the Hasse-Weil zeta function attached to Shimura varieties. In the special case of Shimura varieties classifying polarized abelian varieties with endomorphisms and level structure, Kottwitz established a formula for the number of points with values in a finite field $\mathbb{F}_q$. The formula he obtained is closed to the orbital side of (4) for the reductive group G entering in the definition of $\mathcal{S}$. Again, certain local identities of orbital integrals are needed to establish the equality of number $\sharp\mathcal{S}(\mathbb{F}_q)$ with a combination the orbital sides of (4) for G and a collection of smaller groups called endoscopic groups of G. Eventually, this strategy allows one to attach Galois representation to auto-dual automorphic representations of $\mathrm{GL}(n)$. For the most recent and complete results, see [**31**] and [**38**].

2. Stable trace formula

2.1. Stable conjugacy. In studying orbital integrals for other groups for $\mathrm{GL}(n)$, we encounter with an a priori annoying problem. For $\mathrm{GL}(n)$, two regular semisimple elements in $\mathrm{GL}(n, F)$ are conjugate if and only if they are conjugate in the larger group $\mathrm{GL}(n, \bar{F})$ where $\bar{F}$ is an algebraic closure of F and this latter condition is tantamount to request that γ and γ' have the same characteristic polynomial. For a general reductive group G, we have a characteristic polynomial map $\chi : G \to T/W$ where T is a maximal torus and W is its Weyl group. An element is said strongly regular semisimple if its centralizer is a torus. Strongly regular semisimple elements $\gamma, \gamma' \in G(\bar{F})$ have the same characteristic polynomial if and only if they are $G(\bar{F})$-conjugate. However, there are possibly more than one $G(F)$-conjugacy classes within the set of strongly regular semisimple elements having the same characteristic polynomial in $G(F)$. These conjugacy classes are said stably conjugate.

For a fixed $\gamma \in G(F)$, assumed strongly regular semisimple, the set of $G(F)$-conjugacy classes in the stable conjugacy of γ can be identified with the subset of elements $\mathrm{H}^1(F, I_\gamma)$ whose image in $\mathrm{H}^1(F, G)$ is trivial. For local fields, the group $\mathrm{H}^1(F, I_\gamma)$ is finite but for global field, it can be infinite.

2.2. Stable orbital integral and its κ-sisters. For a local non-archimedean field F, A_γ is a subgroup of the finite abelian group $\mathrm{H}^1(F, I_\gamma)$. One can form linear combinations of orbital integrals within a stable conjugacy class using characters of A_γ. In particular, the stable orbital integral

$$\mathbf{SO}_\gamma(f) = \sum_{\gamma'} \mathbf{O}_{\gamma'}(f)$$

is the sum over a set of representatives γ' of conjugacy classes within the stable conjugacy class of γ. One needs to choose in a consistent way Haar measures on different centralizers $I_{\gamma'}(F)$. For strongly regular semisimple γ, the tori $I_{\gamma'}$ for γ' in the stable conjugacy class of γ, are in fact canonically isomorphic so that we can transfer a Haar measure from $I_\gamma(F)$ to $I_{\gamma'}(F)$.

Obviously, the stable orbital integral $\mathbf{SO}_\gamma$ depends only on the characteristic polynomial of γ. If a is the characteristic polynomial of a strongly regular semisimple element γ, we set $\mathbf{SO}_a = \mathbf{SO}_\gamma$. A stable distribution is an element in the closure of the vector space generated by the distributions of the form $\mathbf{SO}_a$ with respect to the weak topology.

For each character $\kappa : A_\gamma \to \mathbb{C}^\times$, the κ-orbital integral is the linear combination

$$\mathbf{O}_\gamma^\kappa(f) = \sum_{\gamma'} \kappa(\mathrm{cl}(\gamma'))\mathbf{O}_{\gamma'}(f)$$

over a set of representatives γ' of conjugacy classes within the stable conjugacy class of γ, $\mathrm{cl}(\gamma')$ being the class of γ' in A_γ. For any γ' in the stable conjugacy class of γ, A_γ and $A_{\gamma'}$ are canonical isomorphic so that the character κ on A_γ defines a character of A'_γ. Now, $\mathbf{O}_\gamma^\kappa$ and $\mathbf{O}_{\gamma'}^\kappa$ are not equal but differ by the scalar $\kappa(\mathrm{cl}(\gamma'))$ where $\mathrm{cl}(\gamma')$ is the class of γ' in A_γ. Even though this transformation rule is simple enough, we can't a priori define κ-orbital $\mathbf{O}_a^\kappa$ for a characteristic polynomial a as in the case of stable orbital integral. This is a source of an important technical difficulty in the theory of endoscopy that is known as the transfer factor.

At least in the case of Lie algebra, there exists a section $\iota : \mathfrak{t}/W \to \mathfrak{g}$ due to Kostant of the characteristic polynomial map $\chi : \mathfrak{g} \to \mathfrak{t}/W$ and we set

$$\mathbf{O}_a^\kappa = \mathbf{O}_{\iota(a)}^\kappa.$$

Thanks to Kottwitz' calculation of transfer factor, we know that this naively looking definition turns out to be correct. This simplifies significantly the statement of the fundamental lemma and the transfer conjecture for Lie algebra [**22**].

If G is semisimple and simply connected, Steinberg constructed a section $\iota : T/W \to G$ of the characteristic polynomial map $\chi : G \to T/W$. It is tempting to define $\mathbf{O}_a^\kappa$ by Steinberg's section but we don't know whether this is the right definition as in the case of Lie algebra.

2.3. Stabilization. Let F be a global field and $\mathbb{A}$ denote its ring of adeles. Test functions for the trace formula are finite combination of functions f on $G(\mathbb{A})$ of the form $f = \bigotimes_{v\in|F|} f_v$ where for all v, f_v is a smooth function with compact support on $G(F_v)$ and for almost all finite place v, f_v is the characteristic function of $G(\mathcal{O}_v)$ with respect to an integral form of G which is well defined almost everywhere.

The trace formula defines a linear form in f. For each v, it induces an invariant linear form in f_v. There exists a Galois theoretical cohomological obstruction that prevents this linear form from being stably invariant. Let $\gamma \in G(F)$ be a strongly regular semisimple element. Let $(\gamma'_v) \in G(\mathbb{A})$ be an adelic element with γ'_v stably conjugate to γ for all v and conjugate for almost all v. There exists a cohomological obstruction that prevents the adelic conjugacy class (γ'_v) from being rational. In fact the map

$$\mathrm{H}^1(F, I_\gamma) \to \bigoplus_v \mathrm{H}^1(F_v, I_\gamma)$$

is not surjective in general. Let denote $\hat{I}_\gamma$ the dual complex torus of I_γ equipped with a finite action of the Galois group $\Gamma = \mathrm{Gal}(\bar{F}/F)$. For each place v, the Galois group $\Gamma_v = \mathrm{Gal}(\bar{F}_v/F_v)$ of the local field also acts on $\hat{I}_\gamma$. By local Tate-Nakayama duality as reformulated by Kottwitz, $\mathrm{H}^1(F_v, I_\gamma)$ can be identified with the group of characters of $\pi_0(\hat{I}_\gamma^{\Gamma_v})$. By global Tate-Nakayama duality, an adelic class in $\bigoplus_v \mathrm{H}^1(F_v, I_\gamma)$ comes from a rational class in $\mathrm{H}^1(F, I_\gamma)$ if and only if the corresponding characters on $\pi_0(\hat{I}_\gamma^{\Gamma_v})$ restricted to $\pi_0(\hat{I}_\gamma^{\Gamma})$, sum up to the trivial character. The original problem with conjugacy classes within a stable conjugacy class, complicated by the presence of the strict subset A_γ of $\mathrm{H}^1(F, I_\gamma)$, was solved in Langlands [**26**] and in a more general setting by Kottwitz [**20**].

In [**26**], Langlands outlined a program to derive from the usual trace formula a stable trace formula. He proposed to remove first the above Galois cohomological obstruction so that the formula becomes a stable distribution and to introduce the correction terms appearing after a Fourier transform on the obstruction group that similar to the component group $\pi_0(\hat{I}_\gamma^{\Gamma})$. Those correction terms turn out to be κ-orbital integrals. Langlands conjectured that these κ-orbital integrals can also expressed in terms of stable orbital integrals of endoscopic groups. We shall formulate his conjecture with more details later.

Admitting these conjecture on local orbital integrals, Langlands and Kottwitz succeeded to stabilize the elliptic part of the trace formula. In particular, they showed how the different κ-terms for different γ fit in the stable trace formula for endoscopic groups. One of the difficulty is to keep track of the variation of the component group $\pi_0(\hat{I}_\gamma^{\Gamma})$ with γ. The whole trace formula was eventually stabilized by Arthur under the assumption of the weighted fundamental lemma.

In the course of the construction of the stable trace formula, special cases of the functoriality principle between a reductive groups and its endoscopic groups are also established.

2.4. Endoscopic groups. Assume for simplicity that G is a quasi-split group over F that splits over a finite Galois extension K/F. The finite group $\mathrm{Gal}(K/F)$ acts on the root datum of G. Let $\hat{G}$ denote the connected complex reductive group whose root system is related to the root system of G by exchanging roots and coroots. Following [**26**], we set ${}^L G = \hat{G} \rtimes \mathrm{Gal}(K/F)$ where the action of $\mathrm{Gal}(K/F)$ on $\hat{G}$ derives from its action on the root datum. For instant, $G = \mathrm{Sp}(2n)$ and $\hat{G} = \mathrm{SO}(2n+1)$ are dual groups and $\mathrm{SO}(2n)$ is selfdual.

By the Tate-Nakayama duality, a character κ of $\mathrm{H}^1(F, I_\gamma)$ corresponds to a semisimple element $\hat{G}$ well defined up to conjugacy. Let $\hat{H}$ be the neutral component of the centralizer of κ in ${}^L G$. For a given torus I_γ, we can define an action of the Galois group of F on $\hat{H}$ through the component group of the centralizer of κ in ${}^L G$. By duality, we obtain a quasi-split reductive group over F.

More agreeable is the case where the group G is split and has connected centre. In this case, the derived group of $\hat{G}$ is simply connected. This implies that the centralizer $\hat{G}_\kappa$ of the semisimple element κ is connected and therefore the endoscopic group H is split.

2.5. Transfer of stable conjugacy classes. The endoscopic group H is not a subgroup of G in general. Nevertheless, it is possible to transfer stable conjugacy classes from H to G. If G is split and has connected centre, in the dual side $\hat{H} = \hat{G}_\kappa \subset \hat{G}$ induces an inclusion of Weyl groups $W_H \subset W$. It follows that there exists a canonical map $T/W_H \to T/W$ that realizes the transfer of stable conjugacy classes from H to G. If $\gamma_H \in H(F)$ has characteristic polynomial a_H mapping to the characteristic polynomial a of $\gamma \in G(F)$, we will loosely say that γ and γ_H have the same characteristic polynomial.

Similar construction exists for Lie algebras as well. One can transfer stable conjugacy classes in the Lie algebra of H to the Lie algebra of Lie. Moreover, transfer of stable conjugacy classes is not limited to endoscopic relationship. For instant, one can transfer stable conjugacy classes in Lie algebras of groups with isogenous root systems. In particular, this transfer is possible between the Lie algebras of $\mathrm{Sp}(2n)$ and $\mathrm{SO}(2n+1)$.

2.6. Applications of the endoscopy theory. Many known cases of functoriality principle fit in the endoscopic framework. In particular, the transfer known as general Jacquet-Langlands from a group to its quasi-split inner form. The transfer from classical group to $\mathrm{GL}(n)$ expected to follow from Arthur's work on stable trace formula is a particular case of theory of twisted endoscopy.

Endoscopy is however far from exhausting the functoriality principle. It is concerned mainly with "small" homomorphism of L-groups. However, the stable trace formula seems to be an indispensable tool to any serious progress toward understanding functoriality.

Endoscopy is also instrumental in the study of Shimura varieties and the proof of many cases of the global Langlands correspondence [**31**], [**38**].

3. Conjectures on orbital integrals

3.1. Transfer conjecture. The first conjecture is concerned with the possibility of transfer of smooth functions :

CONJECTURE 1. *For every $f \in C_c^\infty(G(F))$ there exists $f^H \in C_c^\infty(H(F))$ such that*

$$\mathbf{SO}_{\gamma_H}(f^H) = \Delta(\gamma_H, \gamma)\mathbf{O}_\gamma^\kappa(f) \tag{5}$$

for all strongly regular semisimple elements γ_H and γ having the same characteristic polynomial, $\Delta(\gamma_H, \gamma)$ being a factor which is independent of f.

Under the assumption γ_H and γ strongly regular semisimple with the same characteristic polynomial, their centralizers in H and G respectively are canonically isomorphic. We can therefore transfer Haar measures between those locally compact groups.

The "transfer" factor $\Delta(\gamma_H, \gamma)$, defined by Langlands and Shelstad in [**27**], is a power of the cardinal of the residue field and a root unity which is a sign in most cases. This sign takes into account the fact that $\mathbf{O}_\gamma^\kappa$ depends on the choice of γ in its stable conjugacy class. In the case of Lie algebra, if we pick $\gamma = \iota(a)$ where ι is the Kostant section to the characteristic polynomial map, this sign equals one, according to Kottwitz in [**22**]. According to Kottwitz again, if the derived group of G is simply connected, Steinberg's section would play the same role for Lie group as Kostant's section for Lie algebra.

3.2. Fundamental lemma. Assume that we are in unramified situation i.e. both G and H have reductive models over $\mathcal{O}_F$. Let $1_{G(\mathcal{O}_F)}$ be the characteristic function of $G(\mathcal{O}_F)$ and $1_{H(\mathcal{O}_F)}$ the characteristic function of $H(\mathcal{O}_F)$.

CONJECTURE 2. *The equality (5) holds for $f = 1_{G(\mathcal{O}_F)}$ and $f^H = 1_{H(\mathcal{O}_F)}$.*

There is a more general version of the fundamental lemma. Let $\mathcal{H}_G$ be the algebra of $G(\mathcal{O}_F)$-biinvariant functions with compact support of $G(F)$ and $\mathcal{H}_H$ the similar algebra for G. Using Satake isomorphism we have a canonical homomorphism $b : \mathcal{H}_G \to \mathcal{H}_H$. Here is the more general version of the fundamental lemma.

CONJECTURE 3. *The equality (5) holds for any $f \in \mathcal{H}_G$ and for $f^H = b(f)$.*

3.3. Lie algebras. There are similar conjectures for Lie algebras. The transfer conjecture can be stated in the same way with $f \in C_c^\infty(\mathfrak{g}(F))$ and $f^H \in C_c^\infty(\mathfrak{h}(F))$. Idem for the fundamental lemma with $f = 1_{\mathfrak{g}(\mathcal{O}_F)}$ and $f^H = 1_{\mathfrak{h}(\mathcal{O}_F)}$.

Waldspurger stated a conjecture called the non standard fundamental lemma. Let G_1 and G_2 be two semisimple groups with isogenous root systems i.e. there exists an isogeny between their maximal tori which maps a root of G_1 on a scalar multiple of a root of G_2 and conversely. In this case, there is an isomorphism $\mathfrak{t}_1/W_1 \simeq \mathfrak{t}_2/W_2$. We can therefore transfer regular semisimple stable conjugacy classes from $\mathfrak{g}_1(F)$ to $\mathfrak{g}_2(F)$ and back.

CONJECTURE 4. *Let $\gamma_1 \in \mathfrak{g}_1(F)$ and $\gamma_2 \in \mathfrak{g}_2(F)$ be regular semisimple elements having the same characteristic polynomial. Then we have*

$$\mathbf{SO}_{\gamma_1}(1_{\mathfrak{g}_1(\mathcal{O}_F)}) = \mathbf{SO}_{\gamma_2}(1_{\mathfrak{g}_2(\mathcal{O}_F)}). \tag{6}$$

The absence of transfer conjecture makes this conjecture particularly agreeable.

3.4. History of the proof. All the above conjectures are now theorems. Let me sketch the contribution of different peoples coming into its proof.

The theory of endoscopy for real groups is almost entirely due to Shelstad.

First case of twisted fundamental lemma was proved by Saito, Shintani and Langlands in the case of base change for GL(2). Kottwitz had a general proof for the fundamental lemma for unit element in the case of base change.

Particular cases of the fundamental lemma were proved by different peoples : Labesse-Langlands for SL(2) [**25**], Kottwitz for SL(3) [**18**], Kazhdan and Waldspurger for SL(n) [**16**], [**39**], Rogawski for U(3) [**4**], Laumon-Ngô for U(n) [**30**], Hales, Schroder and Weissauer for Sp(4). Whitehouse also proved the weighted fundamental lemma for Sp(4).

In a landmark paper, Waldspurger proved that the fundamental lemma implies the transfer conjectures. Due to his and Hales' works, the case of Lie group follows from the case of Lie algera. Waldspurger also proved that the twisted fundamental lemma follows from the combination of the fundamental lemma with his non standard variant [**42**]. In [**13**], Hales proved that if we know the fundamental lemma for the unit for almost all places, we know it for the entire Hecke algebra for all places. In particular, if we know the fundamental lemma for the unit element at all but finitely many places, we also know it at the remaining places.

Following Waldspurger and independently Cluckers, Hales and Loeser, it is enough to prove the fundamental lemma for a local field in characteristic p, see [**41**] and [**6**].

For local fields of Laurent series, the approach using algebraic geometry was eventually successful. The local method was first introduced by Goresky, Kottwitz and MacPherson [**11**] based on the affine Springer fibers constructed by Kazhdan and Lusztig [**17**]. The Hitchin fibration was introduced in this context in [**33**]. Laumon and I used this approach, combined with previous work of Laumon [**29**] in order to prove the fundamental lemma for unitary group in [**30**]. The general case was proved in [**35**] with essentially the same strategy as in [**30**] with one major exception. The equivariant cohomology is no longer used the determination of the support of simple perverse sheaves occurring in the cohomology of Hitchin fibration.

4. Geometric method : local picture

4.1. Affine Springer fibers. Let $k = \mathbb{F}_q$ be a finite field with q elements. Let G be a reductive group over k and $\mathfrak{g}$ its Lie algebra. Let denote

$F = k((\pi))$ and $\mathcal{O}_F = k[[\pi]]$. Let $\gamma \in \mathfrak{g}(F)$ be a regular semisimple element. According to Kazhdan and Lusztig [**17**], there exists a k-scheme $\mathcal{M}_\gamma$ whose set of k points is

$$\mathcal{M}_\gamma(k) = \{g \in G(F)/G(\mathcal{O}_F) \mid \mathrm{ad}(g)^{-1}(\gamma) \in \mathfrak{g}(\mathcal{O}_F)\}.$$

They proved that the affine Springer fiber $\mathcal{M}_\gamma$ is finite dimensional and locally of finite type.

There exists a finite dimensional k-group scheme $\mathcal{P}_\gamma$ acting on $\mathcal{M}_\gamma$. We know that $\mathcal{M}_\gamma$ admits a dense open subset $\mathcal{M}_\gamma^{\mathrm{reg}}$ which is a principal homogenous space of $\mathcal{P}_\gamma$. The group connected components $\pi_0(\mathcal{P}_\gamma)$ of $\mathcal{P}_\gamma$ is possibly infinite and $\mathcal{M}_\gamma$ not of finite type. The group $\mathcal{P}_\gamma$ is a quotient of $I_\gamma(F)$ viewed as infinite dimensional group over k. The action of $\mathcal{P}_\gamma$ on $\mathcal{M}_\gamma$ is induced from $I_\gamma(F)$.

Let us consider a simple but important example. Let $G = \mathrm{SL}_2$ and let γ be the diagonal matrix

$$\gamma = \begin{pmatrix} \pi & 0 \\ 0 & -\pi \end{pmatrix}.$$

In this case $\mathcal{M}_\gamma$ is an infinite chain of projective lines with the point ∞ in each copy being identified with the point 0 of the next one. The group $\mathcal{P}_\gamma$ is $\mathbb{G}_m \times \mathbb{Z}$ with $\mathbb{G}_m$ acts on each copy of $\mathbb{P}^1$ by rescaling and the generator of $\mathbb{Z}$ acts by translation from each copy to the next one. The dense open orbit is obtained by removing from $\mathcal{M}_\gamma$ its double points. The group $\mathcal{P}_\gamma$ over k is closely related to the centralizer of γ is over F which is just the multiplicative group $\mathbb{G}_m$ in this case. The surjective homomorphism

$$I_\gamma(F) = F^\times \to k^\times \times \mathbb{Z} = \mathcal{P}_\gamma(k)$$

attaches to a nonzero Laurent series the first no zero coefficient and the valuation.

In general there is no such an explicit description of the affine Springer fiber. The group $\mathcal{P}_\gamma$ is nevertheless rather explicit. In fact, it can be quite helpful to keep in mind that $\mathcal{M}_\gamma$ is a kind of equivariant compactification of the group $\mathcal{P}_\gamma$.

4.2. Counting points over finite fields. The stabilization of the trace formula suggests that we count the number of points of the quotient $[\mathcal{M}_\gamma/\mathcal{P}_\gamma]$ as an algebraic stack. $[\mathcal{M}_\gamma/\mathcal{P}_\gamma](k)$ is not a set but a groupoid. The cardinal of a groupoid $\mathcal{C}$ is the number

$$\sharp\mathcal{C} = \sum_x \frac{1}{\sharp\mathrm{Aut}(x)}$$

for x in a set of representative of its isomorphism classes and $\sharp\mathrm{Aut}(x)$ being the order of the group of automorphisms of x. In our case, it can be proved that

$$\sharp[\mathcal{M}_\gamma/\mathcal{P}_\gamma](k) = \mathbf{SO}_\gamma(1_{\mathfrak{g}(\mathcal{O}_F)}, dg/dt) \tag{7}$$

for an appropriate choice of Haar measure on the centralizer. Roughly speaking, the Haar measure is related to the kernel of the homomorphism $I_\gamma(F) \to \mathcal{P}_\gamma(k)$.

The group $\pi_0(\mathcal{P}_\gamma)$ of geometric connected components of $\mathcal{P}_\gamma$ is an abelian group of finite type equipped with an action of Frobenius σ_q. For every character of finite order $\kappa : \pi_0(\mathcal{P}_\kappa) \to \mathbb{C}^\times$ fixed by σ_ℓ, we consider the finite sum

$$\sharp[\mathcal{M}_\gamma/\mathcal{P}_\gamma](k)_\kappa = \sum_x \frac{\kappa(\mathrm{cl}(x))}{\sharp\mathrm{Aut}(x)}$$

where $\mathrm{cl}(x) \in \mathrm{H}^1(k, \mathcal{P}_\gamma)$ is the class of the $\mathcal{P}_\gamma$-torsor $\pi^{-1}(x)$ where $\pi : \mathcal{M}_\gamma \to [\mathcal{M}_\gamma/\mathcal{P}_\gamma]$ is the quotient map. By a counting argument similar to the stable case, we have

$$\sharp[\mathcal{M}_\gamma/\mathcal{P}_\gamma](k)_\kappa = \mathbf{O}^\kappa_\gamma(1_{\mathfrak{g}(\mathcal{O}_F)}, dg/dt)$$

This provides a cohomological interpretation for κ-orbital integrals. Let fix an isomorphism $\bar{\mathbb{Q}}_\ell \simeq \mathbb{C}$ so that κ can be seen as taking values in $\bar{\mathbb{Q}}_\ell$. Then we have the formula

$$\mathbf{O}^\kappa_\gamma(1_{\mathfrak{g}(\mathcal{O}_F)}) = \sharp\mathcal{P}^0_\gamma(k)^{-1}\mathrm{tr}(\sigma_q, \mathrm{H}^*(\mathcal{M}_\gamma, \bar{\mathbb{Q}}_\ell)_\kappa).$$

For simplicity, assume that the component group $\pi_0(\mathcal{P}_\gamma)$ is finite. Then $\mathrm{H}^*(\mathcal{M}_\gamma, \bar{\mathbb{Q}}_\ell)_\kappa$ is the biggest direct summand of $\mathrm{H}^*(\mathcal{M}_\gamma, \bar{\mathbb{Q}}_\ell)$ on which $\mathcal{P}_\gamma$ acts through the character κ. When $\pi_0(\mathcal{O}_\gamma)$ is infinite, the definition of $\mathrm{H}^*(\mathcal{M}_\gamma, \bar{\mathbb{Q}}_\ell)_\kappa$ is more complicated.

By taking $\kappa = 1$, we obtained a cohomological interpretation of the stable orbital integral

$$\mathbf{SO}_\gamma(1_{\mathfrak{g}(\mathcal{O}_F)}) = \sharp\mathcal{P}^0_\gamma(k)^{-1}\mathrm{tr}(\sigma_q, \mathrm{H}^*(\mathcal{M}_\gamma, \bar{\mathbb{Q}}_\ell)_{st})$$

where the index *st* means the direct summand where $\mathcal{P}_\gamma$ acts trivially under the assumption $\pi_0(\mathcal{P}_\gamma)$ be finite.

This cohomological interpretation is essentially the same as the one given by Goresky, Kottwitz and MacPherson [**11**]. It allows us to shift focus from a combinatorial problem of counting lattices to a geometric problem of computing ℓ-adic cohomology. However, the calculation of ℓ-adic cohomology of the affine Springer fiber is no easier than the calculation of the orbital integrals themselves.

4.3. More about $\mathcal{P}_\gamma$. We don't know much about $\mathcal{M}_\gamma$. The only information which is available in general is that $\mathcal{M}_\gamma$ is in a loose sense an equivariant compactification of a group $\mathcal{P}_\gamma$ that we know better.

There are two simple but useful facts about the group $\mathcal{P}_\gamma$. A formula for its dimension was conjectured by Kazhdan and Lusztig and proved by Bezrukavnikov [**3**]. The component group $\pi_0(\mathcal{P}_\gamma)$ can also be described precisely. The centralizer I_γ is a torus over F. If G is split, the monodromy of I_γ determines a subgroup $\rho(\Gamma)$ of the Weyl group W well determined up to conjugation. Assume that the center of G is connected. Then $\pi_0(\mathcal{P}_\gamma)$

is the group of $\rho(\Gamma)$-coinvariants of the group of cocharacters $X_*(T)$ of the maximal torus of G. In general, the formula is slightly more complicated.

Let denote $a \in (\mathfrak{t}/W)(F)$ the image of $\gamma \in \mathfrak{g}(F)$. If the affine Springer fiber $\mathcal{M}_\gamma$ is non empty, then a can be extended to a $\mathcal{O}$-point of $\mathfrak{t}/W$. By construction, the group $\mathcal{P}_\gamma$ depends only on $a \in (\mathfrak{t}/W)(\mathcal{O})$ and is denoted by $\mathcal{P}_a$. In general $\mathcal{M}_\gamma$ does not depend only on a. We restrict ourselves to the Kostant section and have an affine Springer fiber $\mathcal{M}_a$ that depends only on a. This choice is consistent with Kottwitz's construction of the transfer factor. This is also helpful for connecting with the global picture.

5. Geometric method : global picture

5.1. The case of SL(2)**.** The description of the Hitchin system in the case of $G = \mathrm{SL}(2)$ is simple and instructive.

Let X be a smooth projective curve over a field k. We assume that X is geometrically connected and its genus is at least 2. A Higgs bundle for SL(2) over X consists in a vector bundle V of rank two with trivialized determinant $\bigwedge^2 V = \mathcal{O}_X$ and equipped with a Higgs field $\phi : V \to V \otimes K$ satisfying the equation $\mathrm{tr}(\phi) = 0$. Here K denotes the canonical bundle and $\mathrm{tr}(\phi) \in \mathrm{H}^0(X, K)$ is a 1-form. The moduli stack of Higgs bundle $\mathcal{M}$ is Artin algebraic and locally of finite type. Let Bun_G denote the moduli stack of principal G-bundles on X. Over the stable locus of Bun_G, $\mathcal{M}$ can be identified with the cotangent of Bun_G by Serre's duality. As a cotangent, $\mathcal{M}$ is naturally equipped with a symplectic structure. Hitchin constructed explicitly a family of d Poisson commuting algebraically independent functions on $\mathcal{M}$ where d is half the dimension of $\mathcal{M}$. In other words, $\mathcal{M}$ is an algebraic completely integrable system.

In SL(2) case, we can associate with a Higgs bundle (V, ϕ) the quadratic differential $a = \det(\phi) \in \mathrm{H}^0(X, K^{\otimes 2})$. By Riemann-Roch, the dimension of $\mathrm{H}^0(X, K^{\otimes 2})$ is also equal to half the dimension of $\mathcal{M}$. According to Hitchin, the mapping $(V, \phi) \mapsto \det(\phi)$ defines a family of d Poisson commuting algebraically independent functions on $\mathcal{M}$.

Following Hitchin, the fibers of the map $f : \mathcal{M} \to \mathcal{A} = \mathrm{H}^0(X, K^{\otimes 2})$ can be described by the recipe of spectral curve. A section $a \in \mathrm{H}^0(X, K^{\otimes 2})$ determines a curve Y_a on the total space $|K|$ of K by pulling back the section $-a$ by the ramified 2-covering $|K| \to |K^{\otimes 2}|$. For any a, $p_a : Y_a \to X$ is a covering of degree 2 of X. If $a \neq 0$, the curve Y_a is reduced. For generic a, the curve Y_a is smooth but in general, it can be singular. It can be even reducible if $a = b^{\otimes 2}$ for certain $b \in \mathrm{H}^0(X, K)$.

By Cayley-Hamilton theorem, if $a \neq 0$, the fiber $\mathcal{M}_a$ can be identified with the moduli space of torsion-free sheaf $\mathcal{F}$ on Y_a such that $\det(p_{a,*}\mathcal{F}) = \mathcal{O}_X$. If Y_a is smooth, $\mathcal{M}_a$ is identified with a translation of a subabelian variety $\mathcal{P}_a$ of the Jacobian of Y_a. This subabelian variety consists in line bundle $\mathcal{L}$ on Y_a such that $\mathrm{Nm}_{Y_a/X}\mathcal{L} = \mathcal{O}_X$.

Hitchin used similar construction of spectral curve to prove that the generic fiber of f is an abelian variety.

5.2. Picard stack of symmetry. Let us observe that the above definition of $\mathcal{P}_a$ is valid for all a. For any a, the group $\mathcal{P}_a$ acts on $\mathcal{M}_a$ because of the formula

$$\det(p_{a,*}(\mathcal{F} \otimes \mathcal{L})) = \det(p_{a,*}\mathcal{F}) \otimes \mathrm{Nm}_{Y_a/X}\mathcal{L}.$$

In [**33**], we construct $\mathcal{P}_a$ and its action on $\mathcal{M}_a$ for any reductive group. Instead of the canonical bundle, K can be any line bundle of large degree. We defined a canonical Picard stack $g : \mathcal{P} \to \mathcal{A}$ acting on the Hitchin fibration $f : \mathcal{M} \to \mathcal{A}$ relatively to the base $\mathcal{A}$. In general, $\mathcal{P}_a$ does not act simply transitively on $\mathcal{M}_a$. It does however on a dense open subset of $\mathcal{M}_a$. This is why we can think about the Hitchin fibration $\mathcal{M} \to \mathcal{A}$ as an equivariant compactification of the Picard stack $\mathcal{P} \to \mathcal{A}$.

Consider the quotient $[\mathcal{M}_a/\mathcal{P}_a]$ of the Hitchin fiber $\mathcal{M}_a$ by its natural group of symmetries. In [**33**], we observed a product formula

$$[\mathcal{M}_a/\mathcal{P}_a] \simeq \prod_v [\mathcal{M}_{v,a}/\mathcal{P}_{v,a}] \tag{8}$$

where for all $v \in X$, $\mathcal{M}_{v,a}$ is the affine Springer fiber at the place v attached to a and $\mathcal{P}_a$ is its symmetry group that appeared in 4.3. For all but finitely many v, $\mathcal{P}_{a,v}$ acts simply transitively on $\mathcal{M}_{v,a}$. The sign $\simeq$ means homeomorphism. It does not seem to be an isomorphism in general. However, for the purpose of ℓ-adic cohomology, it does not make any difference with an isomorphism.

Even though the Hitchin fibers $\mathcal{M}_a$ are organized in a family, individually, their structure depends significantly on a. For generic a, $\mathcal{P}_a$ acts simply transitively on $\mathcal{M}_a$ so that all quotients appearing in the product formula are trivial. In this case, all affine Springer fibers appearing on the right hand side are zero dimensional. For bad parameter a, affine Springer fibers have positive dimension. The Hitchin fibration allows us to have a control on the bad fibers from the good fibers. This is the basic idea of our global geometric method.

5.3. Counting points with values in a finite field. Let k be a finite field of characteristic p with q elements. In counting the numbers of points with values in k on a Hitchin fiber, we noticed a remarkable connection with the trace formula.

In choosing a global section of K, we identify K with the line bundle $\mathcal{O}_X(D)$ attached to an effective divisor D. It also follows an injective map $a \mapsto a_F$ from $\mathcal{A}(k)$ into $(\mathfrak{t}/W)(F)$. The image is a finite subset of $(\mathfrak{t}/W)(F)$ that can be described easily with help of the exponents of $\mathfrak{g}$ and the divisor D. Thus points on the Hitchin base correspond essential to rational stable conjugacy classes, see [**33**] and [**34**].

For simplicity, assume that the kernel $\ker^1(F,G)$ of the map

$$\mathrm{H}^1(F,G) \to \prod_v \mathrm{H}^1(F_v,G)$$

is trivial. Following Weil's adelic desription of vector bundle on a curve, we can express the number of points on $\mathcal{M}_a = f^{-1}(a)$ as a sum of global orbital integrals

$$\sharp\mathcal{M}_a(k) = \sum_\gamma \int_{I_\gamma(F)\backslash G(\mathbb{A}_F)} 1_D(\mathrm{ad}(g)^{-1}\gamma)dg \tag{9}$$

where γ runs over the set of conjugacy classes of $\mathfrak{g}(F)$ with a as the characteristic polynomial, F being the field of rational functions on X, $\mathbb{A}_F$ the ring of adèles of F, 1_D a very simple function on $\mathfrak{g}(\mathbb{A}_F)$ associated with a choice of divisor within the linear equivalence class D. In summing over $a \in \mathcal{A}(k)$, we get an expression very similar to the geometric side of the trace formula for Lie algebra.

Without the assumption on the triviality of $\ker^1(F,G)$, we obtain a sum of trace formula for inner form of G induced by elements of $\ker^1(F,G)$. This further complication turns out to be a simplification when we stabilize the formula, see [**34**]. In particular, instead of the subgroup A_γ of $\mathrm{H}^1(F,I_\gamma)$ as in 2.1, we deal with the group $\mathrm{H}^1(F,I_\gamma)$ it self.

At this point, it is a natural to seek a geometric interpretation of the stabilization process as explained in 2.3. Fix a rational point $a \in \mathcal{A}(k)$ and consider the quotient morphism

$$\mathcal{M}_a \to [\mathcal{M}_a/\mathcal{P}_a]$$

If $\mathcal{P}_a$ is connected then for every point $x \in [\mathcal{M}_a/\mathcal{P}_a](k)$, there is exactly $\sharp\mathcal{P}_a(k)$ points with values in k in the fiber over x. It follows that

$$\sharp\mathcal{M}_a(k) = \sharp\mathcal{P}_a(k)\sharp[\mathcal{M}_a/\mathcal{P}_a](k)$$

where $\sharp[\mathcal{M}_a/\mathcal{P}_a](k)$ can be expressed by stable orbital integrals by the product formula 8 and by 7. In general, the component group $\pi_0(\mathcal{P}_a)$ prevents the number $\sharp\mathcal{M}_a(k)$ from being expressed as stable orbital integrals.

5.4. Variation of the component groups $\pi_0(\mathcal{P}_a)$. The dependence of the component group $\pi_0(\mathcal{P}_a)$ on a makes the combinatorics of the stabilization of the trace formula rather intricate. Geometrically, this variation can be packaged in a sheaf of abelian group $\pi_0(\mathcal{P}/\mathcal{A})$ over $\mathcal{A}$ whose fibers are $\pi_0(\mathcal{P}_a)$.

If the center G is connected, it is not difficult to express $\pi_0(\mathcal{P}_a)$ from a in using a result of Kottwitz [**21**]. A point $a \in \mathcal{A}(\bar{k})$ defines a stable conjugacy class $a_F \in (\mathfrak{t}/W)(F \otimes_k \bar{k})$. We assume a_F is regular semisimple so that there exists $g \in \mathfrak{g}(F \otimes_k \bar{k})$ whose characteristic polynomial is a. The centralizer I_x is a torus which does not depend on the choice of x but only on a. Its monodromy can expressed as a homomorphism $\rho_a : \mathrm{Gal}(F \otimes_k \bar{k}) \to \mathrm{Aut}(\mathbb{X}_*)$ where $\mathbb{X}_*$ is the group of cocharacters of a maximal torus of G.

The component group $\pi_0(\mathcal{P}_a)$ is isomorphic to the group of coinvariants of $\mathbb{X}_*$ under the action of $\rho_a(\mathrm{Gal}(F \otimes_k \bar{k}))$.

This isomorphism can be made canonical after choosing a rigidification. Let's fix a point $\infty \in X$ and choose a section of the line bundle K non vanishing on a neighborhood of ∞. Consider the covering $\tilde{\mathcal{A}}$ of $\mathcal{A}$ consisting of a pair $\tilde{a} = (a, \tilde{\infty})$ tale where $a \in \mathcal{A}$ regular semisimple at ∞ i.e. $a(\infty) \in (\mathfrak{t}/W)^{rs}$ and $\tilde{\infty} \in \mathfrak{t}^{rs}$ mapping to $a(\infty)$. The map $\tilde{\mathcal{A}} \to \mathcal{A}$ is etale, more precisely, finite etale over a Zariski open subset of $\mathcal{A}$. Over $\tilde{\mathcal{A}}$, there exists a surjective homomorphism from the constant sheaf $\mathbb{X}_*$ to $\pi_0(\mathcal{P})$ whose fiber admits now a canonical description as coinvariants of $\mathbb{X}_*$ under certain subgroup of the Weyl group depending on a.

When the center of G isn't connected, the answer is somehow subtler. In the SL_2 case, there are three possibilities. We say that a is generic, or stable if the spectral curve Y_a has at least one unibranched ramification point over X. In particular, if Y_a is smooth, all ramification points are unibranched. In this case $\pi_0(\mathcal{P}_a) = 0$. We say that a is hyperbolic if the spectral curve Y_a is reducible. In this case on can express $a = b^{\otimes 2}$ for some $b \in \mathrm{H}^0(X, K)$. If a is hyperbolic, we have $\pi_0(\mathcal{P}_a) = \mathbb{Z}$. The most interesting case is the case where a is neither stable nor hyperbolic i.e. the spectral curve Y_a is irreducible but all ramification points have two branches. In this case $\pi_0(\mathcal{P}_a) = \mathbb{Z}/2\mathbb{Z}$ and we say that a is endoscopic. We observe that a is endoscopic if and only if the normalization of Y_a is an unramified double covering of X. Such a covering corresponds to a nontrivial line bundle $\mathcal{E}$ on X such that $\mathcal{E}^{\otimes 2} = \mathcal{O}_X$. Moreover we can express $a = b^{\otimes 2}$ where $b \in \mathrm{H}^0(X, K \otimes \mathcal{E})$.

The upshot of this calculation can be summarized as follows. The free rank of $\pi_0(\mathcal{P}_a)$ jumps exactly when a is hyperbolic i.e. when a comes from a Levi subgroup of G. The torsion group of $\pi_0(\mathcal{P}_a)$ jumps exactly when a is endoscopic i.e. when a comes from an endoscopic group of G. These statement are in fact valid in general.

5.5. Stable part. We can construct an open subset $\mathcal{A}^{ani}$ of $\mathcal{A}$ over which $\mathcal{M} \to \mathcal{A}$ is proper and $\mathcal{P} \to \mathcal{A}$ is of finite type. In particular for every $a \in \mathcal{A}^{ani}(\bar{k})$, the component group $\pi_0(\mathcal{P}_a)$ is a finite group. In fact the converse assertion is also true : $\mathcal{A}^{ani}$ is precisely the open subset of $\mathcal{A}$ where the sheaf $\pi_0(\mathcal{P}/\mathcal{A})$ is an finite.

By construction, $\mathcal{P}$ acts on direct image $f_*\mathbb{Q}_\ell$ as an oject of the derived category of ℓ-adic sheaves on $\mathcal{A}$. The homotopy lemma implies that the induced action on the perverse sheaves of cohomology ${}^p\mathrm{H}^n(f_*\mathbb{Q}_\ell)$ factors through the sheaf of components $\pi_0(\mathcal{P}/\mathcal{A})$ which is finite over $\mathcal{A}^{ani}$. Over this open subset, Deligne's theorem assures the purity of the above perverse sheaves. The finite action of $\pi_0(\mathcal{P}/\mathcal{A}^{ani})$ decomposes ${}^p\mathrm{H}^n(f_*^{ani}\mathbb{Q}_\ell)$ into a direct sum.

This decomposition is at least as complicated as the sheaf $\pi_0(\mathcal{P}/\mathcal{A})$. In fact, this reflects exactly the combinatoric complexity of the stabilization of the trace formula as we have seen in 2.3.

We define the stable part ${}^p\mathrm{H}^n(f_*^{ani}\mathbb{Q}_\ell)_{st}$ as the largest direct factor acted on trivially by $\pi_0(\mathcal{P}/\mathcal{A}^{ani})$. For every $a \in \mathcal{A}^{ani}(k)$, it can be showed by using the argument of 5.3 that the alternating sum of the traces of the Frobenius operator σ_a on ${}^p\mathrm{H}^n(f_*\mathbb{Q}_\ell)_{st,a}$ can be expressed as stable orbital integrals.

THEOREM 1. *Assume $k = \mathbb{C}$. For every integer n the perverse sheaf ${}^p\mathrm{H}^i(f_*^{ani}\mathbb{Q}_\ell)_{st}$ is completely determined by its restriction to any non empty open subset of $\mathcal{A}$. More preceisely, it can be recovered from its restriction by the functor of intermediate extension.*

When k is a finite field, we proved a weaker variant of this theorem which is strong enough the proof of the fundamental lemma. For simplicity, let's pretend here that the above theorem is also proved in positive characteristic.

Let G_1 and G_2 be two semisimple groups with isogenous root systems like $\mathrm{Sp}(2n)$ and $\mathrm{SO}(2n+1)$. The corresponding Hitchin fibration $f_\alpha : \mathcal{M}_\alpha \to \mathcal{A}$ for $\alpha \in \{1, 2\}$ map to the same base. For a generic a, $\mathcal{P}_{1,a}$, and $\mathcal{P}_{2,a}$ are essentially isogenous abelian varieties. It follows that ${}^p\mathrm{H}^i(f_{1,*}\mathbb{Q}_\ell)_{st}$ and ${}^p\mathrm{H}^i(f_{2,*}\mathbb{Q}_\ell)_{st}$ restricted to a non empty open subset of $\mathcal{A}$ are isomorphic local systems. With the intermediate extension, we obtain an isomorphism between perverse sheaves ${}^p\mathrm{H}^i(f_{1,*}\mathbb{Q}_\ell)_{st}$ and ${}^p\mathrm{H}^i(f_{2,*}\mathbb{Q}_\ell)_{st}$. We derive from this isomorphism the Waldspurger conjecture 6.

5.6. Support. By decomposition theorem, the pure perverse sheaves ${}^p\mathrm{H}^n(f_*^{ani}\mathbb{Q}_\ell)$ are geometrically direct sum of simple perverse sheaves. Following Goresky and MacPherson, for a simple perverse sheaf K over base S, there exists an irreducible closed subscheme $i : Z \hookrightarrow S$ of S, an open subscheme $j : U \hookrightarrow Z$ of Z and a local system $\mathcal{K}$ on Z such that $K = i_* j_{!*}\mathcal{K}[\dim(Z)]$. In particular, the support $Z = \mathrm{supp}(K)$ is well defined.

The theorem 1 can be reformulated as follows. Let K be a simple perverse sheaf geometric direct factor of ${}^p\mathrm{H}^i(f_*^{ani}\mathbb{Q}_\ell)_{st}$. Then the support of K is the whole base $\mathcal{A}$.

In general, the determination of the support of constituents of a direct image is a rather difficult problem. This problem is solved to a large extent for Hitchin fibration and for more general abelian fibration. The complete answer involves endoscopic parts as well as the stable part.

5.7. Endoscopic part. Consider again the SL_2 case. In this case $\mathcal{A} - \{0\}$ is the union of closed strata $\mathcal{A}^{hyp}$ and $\mathcal{A}^{endo}$ that are the hyperbolic and endoscopic loci and the open stratum $\mathcal{A}^{st}$. The anisotropic open subset is $\mathcal{A}^{endo} \cup \mathcal{A}^{st}$. Over $\mathcal{A}^{ani}$, the sheaf $\pi_0(\mathcal{P})$ is the unique quotient of the constant sheaf $\mathbb{Z}/2\mathbb{Z}$ that is trivial on the open subset $\mathcal{A}^{st}$ and non trivial on the closed subset $\mathcal{A}^{endo}$.

The group $\mathbb{Z}/2\mathbb{Z}$ acts on ${}^p\mathrm{H}^n(f_*^{ani}\mathbb{Q}_\ell)$ and decomposes it into a direct sum

$$ {}^p\mathrm{H}^n(f_*^{ani}\mathbb{Q}_\ell) = {}^p\mathrm{H}^n(f_*^{ani}\mathbb{Q}_\ell)_+ \oplus {}^p\mathrm{H}^n(f_*^{ani}\mathbb{Q}_\ell)_-. $$

By its very construction, the restriction of the odd part ${}^p\mathrm{H}^n(f_*^{ani}\mathbb{Q}_\ell)_-$ to the open subset $\mathcal{A}^{st}$ is trivial.

For every simple perverse sheaf K direct factor of ${}^p\mathrm{H}^n(f_*^{ani}\mathbb{Q}_\ell)_-$, the support of K is contained in one of the irreducible components of the endoscopic locus $\mathcal{A}^{endo}$. In reality, we prove that the support of a simple perverse sheaf K direct factor of ${}^p\mathrm{H}^n(f_*^{ani}\mathbb{Q}_\ell)_-$ is one of the irreducible components of the endoscopic locus.

In general case, the monodromy of $\pi_0(\mathcal{P}/\mathcal{A})$ prevents the result from being formulated in an agreeable way. We encounter again with the complicated combinatoric in the stabilization of the trace formula. In geometry, it is possible to avoid this unpleasant combinatoric by passing to the etale covering $\tilde{\mathcal{A}}$ of $\mathcal{A}$ defined in 5.4. Over $\tilde{\mathcal{A}}$, we have a surjective homomorphism from the constant sheaf $\mathbb{X}_*$ onto the sheaf of component group $\pi_0(\mathcal{P}/Ac)$ which is finite over $\mathcal{A}^{ani}$. Over $\mathcal{A}^{ani}$, there is a decomposition in direct sum

$$ {}^p\mathrm{H}^n(\tilde{f}_*^{ani}\mathbb{Q}_\ell) = \bigoplus_\kappa {}^p\mathrm{H}^n(\tilde{f}_*^{ani}\mathbb{Q}_\ell)_\kappa $$

where $\tilde{f}^{ani}$ is the base change of f to $\tilde{\mathcal{A}}^{ani}$ and κ are characters of finite order $\mathbb{X}_* \to \mathbb{Q}_\ell{}^\times$.

For any κ as above, the set of geometric points $\tilde{a} \in \tilde{\mathcal{A}}^{ani}$ such that κ factors through $\pi_0(\mathcal{P}_a)$, forms a closed subscheme $\tilde{\mathcal{A}}^{ani}_\kappa$ of $\tilde{\mathcal{A}}^{ani}$. One can check that the connected components of $\tilde{\mathcal{A}}^{ani}_\kappa$ are exactly of the form $\tilde{\mathcal{A}}^{ani}_H$ for endoscopic groups H that are certain quasi-split groups with $\hat{H} = \hat{G}^0_\kappa$.

THEOREM 2. *Let $k = \mathbb{C}$. Let K be a simple perverse sheaf geometric direct factor of $\tilde{\mathcal{A}}^{ani}_\kappa$. Then the support of K is one of the $\tilde{\mathcal{A}}_H$ as above.*

Again, in characteristic p, we prove a weaker form which is strong enough to imply the fundamental lemma. Let's pretend here that the above theorem is proved in positive characteristic.

The geometric version of the fundamental lemma states that the restriction of ${}^p\mathrm{H}^n(\tilde{f}_*^{ani}\mathbb{Q}_\ell)_\kappa$ to $\tilde{\mathcal{A}}_H$ is isomorphic with ${}^p\mathrm{H}^{n+2r}(\tilde{f}_{H,*}^{ani}\mathbb{Q}_\ell)_{st}(-r)$ for certain shifting integer r. Here f_H is the Hitchin fibration for H and $\tilde{f}_H^{ani}$ is its base change to $\tilde{\mathcal{A}}_H^{ani}$. The support theorems 1 and 2 allow us to reduce the problem to an arbitrarily small open subset of $\tilde{\mathcal{A}}_H^{ani}$. On a small open subset of $\tilde{\mathcal{A}}_H^{ani}$, this isomorphism can be constructed by direct calculation, mainly based on the example of the infinite chain of projective lines as in the case of SL(2).

We refer to **[37]** for an account of the proof of the support theorem.

6. Weighted fundamental lemma

In order to stabilize the whole trace formula, Arthur needs more complicated local identities known as weighted fundamental lemma. These identities, conjectured by Arthur, are now theorems due to efforts of Chaudouard, Laumon and Waldspurger. As in the case of the fundamental lemma, Waldspurger proved that the weighted fundamental lemma for a p-adic field is equivalent to the same lemma for the Laurent formal series field $\mathbb{F}_p((\pi))$ as long as the residual characteristic is large with respect to the group G. Chaudouard, Laumon also used the Hitchin fibration and a support theorem to prove the weighted fundamental lemma in positive characteristic case.

The weighted fundamental lemma as stated by Arthur is rather intricate a combinatorial identity. It is in fact easier to explain the weighted fundamental lemma from the point of view of the Hitchin fibration than from the point of view of the trace formula.

We already observed that over the open subset $\mathcal{A}^{ani}$ of $\mathcal{A}$, the Hitchin fibration $f^{ani} : \mathcal{M}^{ani} \to \mathcal{A}^{ani}$ is a proper map. Chaudouard and Laumon made the important observation that an appropriate stability condition make it possible to extend f^{ani} to a proper map $f^{\chi-st} : \mathcal{M}^{\chi-st} \to \mathcal{A}^{\heartsuit}$ where $\mathcal{A}^{\heartsuit}$ is the open subset of $\mathcal{A}$ consisting in $a \in \mathcal{A}$ with regular semisimple generic fiber $a_F \in (\mathfrak{t}/W)(F \otimes_k \bar{k})$.

The stability condition depends on an arbitrary choice of $\chi \in \mathbb{X}_* \otimes \mathbb{R}$. For general χ, the condition χ-stability and χ-semistability become equivalent. For those χ, the morphism $f^{\chi-st} : \mathcal{M}^{\chi-st} \to \mathcal{A}^{\heartsuit}$ is proper. In counting number of points on the fibers of $f^{\chi-st}$, they obtained formula involving weighted orbital integrals. Remarkably, this formula shows that the number of points does not depends on the choice of χ. Chaudouard and Laumon were also able to extend the support theorems 1 and 2 and from this deduce the weighted fundamental lemma [**5**].

7. Perspective

The method used to prove the fundamental lemma should be useful to local identities issued from the comparison of trace formula and relative trace formulas. In fact the first instance of fundamental lemma proved by this geometric method is a relative fundamental lemma conjectured by Jacquet and Ye [**32**]. Recently, Z. Yun proved a fundamental lemma conjectured by Jacquet, Rallis [**43**]. We can expect that other fundamental lemmas can be proved following the same general pattern too. Technically, it can still be challenging. In fact, the support theorem was proved by three completely different method in each of the three cases Jacquet-Ye, Langlands-Shelstad or Jacquet-Rallis. In the unitary case, a weak version of the support theorem was proved by yet another method by Laumon and myself.

The general method is based so far on a geometric interpretation of the orbital side of the trace formula. It is legitimate to ask if it is possible to insert geometry to the spectral side as well. At least for a Riemann surface, the answer seems to be yes. In a joint work in progress with E. Frenkel and R. Langlands, we noticed a closed relationship between the trace formula and Beilinson-Drinfeld's conjecture in geometric Langlands program. We should mention the related work [**10**] of Frenkel and Witten on a manifestation of endoscopy in Kapustin-Witten's proposal for geometric Langlands conjecture.

The endoscopy theory has been essentially completed. We have at our disposal the stable trace formula. It seems now the great times to read "Beyond endoscopy" written by Langlands some years ago [**28**]. Though the difficulty is formidable, his proposal possibly leads us to the understanding of the functoriality of automorphic representations.

References

[1] Arthur J. An introduction to the trace formula. Harmonic analysis, the trace formula, and Shimura varieties, 1–263, Clay Math. Proc., 4, Amer. Math. Soc., Providence, RI, (2005).

[2] Beilinson A., Bernstein J., Deligne P. Faisceaux pervers. *Astérisquc* 100 (1982).

[3] Bezrukavnikov R. The dimension of the fixed point set on affine flag manifolds. Math. Res. Lett. 3 (1996), no. 2, 185–189.

[4] Blasius D., Rogawski J. Fundamental lemmas for U(3) and related groups. in The zeta functions of Picard modular surfaces, 363–394, Univ. Montral, Montreal, QC, (1992).

[5] Chaudouard, P.-H., Laumon, G. Le lemme fondamental pondéré I et II. Preprints.

[6] Cluckers R., Hales T., Loeser F. Transfer principle for the fundamental lemma, preprint.

[7] Dat, J.-F. Lemme fondamental et endoscopie, une approche géométrique, *Séminaire Bourbaki* 940 novembre 2004.

[8] Dat, J.-F. Ngo Dac, T. Le lemme fondamental pour les algèbres de Lie d'après Ngô Bao Chau, Book Project.

[9] Deligne P. La conjecture de Weil II. *Publ. Math. de l'I.H.E.S.* 52 (1980) 137–252.

[10] Frenkel E. , Witten, E. Geometric endoscopy and mirror symmetry. Commun. Number Theory Phys. 2 (2008), no. 1, 113–283.

[11] Goresky M., Kottwitz R., MacPherson R. Homology of affine Springer fiber in the unramified case. *Duke Math. J.* 121 (2004) 509–561.

[12] Goresky M., Kottwitz R., MacPherson R. Codimensions of root valuation strata. Preprint.

[13] Hales T. On the fundamental lemma for standard endoscopy: reduction to unit elements. *Canad. J. Math.* 47 (1995), no. 5, 974–994.

[14] Harris, M. Introduction in Book project.

[15] Hitchin N. Stable bundles and integrable connections. *Duke Math. J.* 54 (1987) 91–114.

[16] Kazhdan D. On liftings. In Lie group representations, II (College Park, Md., 1982/1983), 209–249,

[17] Kazhdan D., Lusztig G. Fixed point varieties on affine flag manifolds. *Israel J. Math.* 62 (1988), no. 2, 129–168.

[18] Kottwitz, R. Unstable orbital integrals on SL(3). Duke Math. J. 48 (1981), no. 3, 649–664.

[19] Kottwitz R. Stable trace formula: cuspidal tempered terms. Duke Math J. (1984) vol. 51 (3) pp. 611–650.

[20] Kottwitz R. Stable trace formula: elliptic singular terms. Math. Ann. 275 (1986), no. 3, 365–399.

[21] Kottwits R. Isocrystals with additional structure. Compositio Math. 56 (1985), no. 2, 201–220.

[22] Kotttwiz R. Transfert factors for Lie algebra. *Represent. Theory* 3 (1999) 127–138.
[23] Kottwitz R. Shimura varieties and λ-adic representations, in *Automorphic forms, Shimura varieties, and L-functions*, Vol. I 161–209, Perspect. Math., 10, Academic Press, Boston, MA, (1990).
[24] Labesse, J.-P. Introduction to endoscopy. In Representation theory of real reductive Lie groups, 175–213, Contemp. Math., 472, Amer. Math. Soc., Providence, RI, (2008).
[25] Labesse, J.-P. Langlands, R. L-indistinguishability for SL(2). Canad. J. Math. 31 (1979), no. 4, 726–785.
[26] Langlands R. *Les débuts d'une formule des traces stables.* Publications de l'Université Paris 7, 13 (1983).
[27] Langlands R., Shelstad D. On the definition of transfer factors. *Math. Ann.* 278 (1987), no. 1–4, 219–271.
[28] Langlands R. Beyond endoscopy. The version posted on the website http://publications.ias.edu/rpl/ is preferred.
[29] Laumon, G. Sur le lemme fondamental pour les groupes unitaires. Preprint.
[30] Laumon, G. et Ngô B.C.: Le lemme fondamental pour les groupes unitaires, *Annals of Math.* 168 (2008), no. 2, 477–573.
[31] Morel, S. On the cohomology of certain non-compact Shimura varieties. Preprint.
[32] Ngô B.C. Le lemme fondamental de Jacquet et Ye en caractristique positive. Duke Math. J. 96 (1999), no. 3, 473–520.
[33] Ngô B.C. Fibration de Hitchin et endoscopie. *Inv. Math.* 164 (2006) 399–453.
[34] Ngô B.C. Fibration de Hitchin et structure endoscopique de la formule des traces. International Congress of Mathematicians. Vol. II, 1213–1225, Eur. Math. Soc., Zrich, 2006.
[35] Ngô B.C. Le lemme fondamental pour les algèbres de Lie. Preprint.
[36] Ngô B.C. Vietnamese congress of mathematicians (2008).
[37] Ngô B.C. Decomposition theorem and abelian fibration, Book project.
[38] Shin S.-W. Galois representations arising from some compact Shimura varieties. Preprint.
[39] Waldspurger, J.-L. Sur les intgrales orbitales tordues pour les groupes linaires: un lemme fondamental. Canad. J. Math. 43 (1991), no. 4, 852–896.
[40] Waldspurger, J.-L. Le lemme fondamental implique le transfert. Compositio Math. 105 (1997), no. 2, 153–236.
[41] Waldspurger J.-L. Endoscopie et changement de caractéristique, J. Inst. Math. Jussieu 5 (2006), no. 3, 423–525.
[42] Waldspurger J.-L. L'endoscopie tordue n'est pas si tordue. Mem. Amer. Math. Soc. 194 (2008), no. 908, x+261 pp.
[43] Yun Z. The fundamental lemma of Jacquet-Rallis in positive characteristics. Preprint.

School of mathematics, Institute for Advanced Study, Princeton NJ 08540 USA, and Département de mathématiques, Université Paris-Sud, 91405 Orsay France

Drawing by Carolyn Snaith, 1981

The Arf-Kervaire Invariant Problem in Algebraic Topology: Introduction

Michael A. Hill, Michael J. Hopkins, and Douglas C. Ravenel

ABSTRACT. This paper gives the history and background of one of the oldest problems in algebraic topology, along with a short summary of our solution to it and a description of some of the tools we use. More details of the proof are provided in our second paper in this volume, *The Arf-Kervaire invariant problem in algebraic topology: Sketch of the proof*. A rigorous account can be found in our preprint *The non-existence of elements of Kervaire invariant one* on the arXiv and on the third author's home page. The latter also has numerous links to related papers and talks we have given on the subject since announcing our result in April, 2009.

CONTENTS

1. Background and history 26
1.1. Pontryagin's early work on homotopy groups of spheres 26
1.2. Our main result 32
1.3. The manifold formulation 33
1.4. The unstable formulation 36
1.5. Questions raised by our theorem 39
2. Our strategy 39
2.1. Ingredients of the proof 39
2.2. The spectrum Ω 40
2.3. How we construct Ω 40
3. Some classical algebraic topology 41
3.1. Fibrations 41
3.2. Cofibrations 42
3.3. Eilenberg-Mac Lane spaces and cohomology operations 44
3.4. The Steenrod algebra. 45
3.5. Milnor's formulation 46
3.6. Serre's method of computing homotopy groups 47
3.7. The Adams spectral sequence 48
4. Spectra and equivariant spectra 50
4.1. An informal definition of spectra 50
4.2. Orthogonal spectra 51
4.3. Equivariant orthogonal spectra 52
References 55

MAIN THEOREM. *The Arf-Kervaire elements $\theta_j \in \pi^S_{2^{j+1}-2}$ do not exist for $j \geq 7$.*

Here π^S_k denotes the kth stable homotopy group of spheres, which will be defined shortly.

The kth (for a positive integer k) homotopy group of the topological space X, denoted by $\pi_k(X)$, is the set of continuous maps to X from the k-sphere S^k, up to continuous deformation. For technical reasons we require that each map send a specified point in S^k (called a *base point*) to a specified point $x_0 \in X$. When X is path connected the choice of these two points is irrelevant, so it is usually omitted from the notation. When X is not path connected, we get different collections of maps depending on the path connected component of the base point.

This set has a natural group structure, which is abelian for $k > 1$. The word *natural* here means that a continuous base point preserving map $f : X \to Y$ induces a homomorphism $f_* : \pi_k(X) \to \pi_k(Y)$, sometimes denoted by $\pi_k(f)$.

It is known that the group $\pi_{n+k}(S^n)$ is independent of n for $n > k$. There is a homomorphism $E : \pi_{n+k}(S^n) \to \pi_{n+k+1}(S^{n+1})$ defined as follows. S^{n+1} $[S^{n+k+1}]$ can be obtained from S^n $[S^{n+k}]$ by a double cone construction known as *suspension*. The cone over S^n is an $(n+1)$-dimensional ball, and gluing two such balls together along there common boundary gives and $(n+1)$-dimensional sphere. A map $f : S^{n+k} \to S^n$ can be canonically extended (by suspending both its source and target) to a map $Ef : S^{n+k+1} \to S^{n+1}$, and this leads to the suspension homomorphism E. The Freudenthal Suspension Theorem [**Fre38**], proved in 1938, says that it is onto for $k = n$ and an isomorphism for $n > k$. For this reason the group $\pi_{n+k}(S^n)$ is said to be *stable* when $n > k$, and it is denoted by π_k^S and called the *stable* k*-stem*.

The Main Theorem above concerns the case $k = 2^{j+1} - 2$. The θ_j in the theorem is a hypothetical element related a geometric invariant of certain manifolds studied originally by Pontryagin starting in the 1930s, [**Pon38**], [**Pon50**] and [**Pon55**]. The problem came into its present form with a theorem of Browder [**Bro69**] published in 1969. There were several unsuccessful attempts to solve it in the 1970s. They were all aimed at proving the opposite of what we have proved, namely that ll of the θ_j exist.

The θ_j in the theorem is the name given to a hypothetical map between spheres for which the Arf-Kervaire invariant is nontrivial. Browder's theorem says that such things can exist only in dimensions that are 2 less than a power of 2.

Some homotopy theorists, most notably Mahowald, speculated about what would happen if θ_j existed for all j. They derived numerous consequences about homotopy groups of spheres. The possible nonexistence of the θ_j for large j was known as the DOOMSDAY HYPOTHESIS.

After 1980, the problem faded into the background because it was thought to be too hard. In 2009, just a few weeks before we announced our theorm, Snaith published a book [**Sna09**] on the problem "to stem the tide of oblivion". On the difficulty of the problem, he wrote

> In the light of ... the failure over fifty years to construct framed manifolds of Arf-Kervaire invariant one this might turn out to be a book about things which do not exist. This [is] why the quotations which preface each chapter contain a preponderance of utterances from the pen of Lewis Carroll.

Our proof is two giant steps away from anything that was attempted in the 70s. We now know that the world of homotopy theory is very different from what they had envisioned then.

1. Background and history

1.1. Pontryagin's early work on homotopy groups of spheres. The Arf-Kervaire invariant problem has its origins in Pontryagin's early work on a geometric approach to the homotopy groups of spheres, [**Pon38**], [**Pon50**] and [**Pon55**].

Pontryagin's approach to maps $f : S^{n+k} \to S^n$ is to assume that f is smooth and that the base point y_0 of the target is a regular value. (Any continuous f can be continuously deformed to a map with this property.) This means that $f^{-1}(y_0)$ is a closed smooth k-manifold M in S^{n+k}. Let D^n be the closure of an open ball around y_0. If it is sufficiently small, then $V^{n+k} = f^{-1}(D^n) \subset S^{n+k}$ is an $(n+k)$-manifold homeomorphic to $M \times D^n$ with boundary homeomorphic to $M \times S^{n-1}$. It is also a tubular neighborhood of M^k and comes equipped with a map $p : V^{n+k} \to M^k$ sending each point to the nearest point in M. For each $x \in M$, $p^{-1}(x)$ is homeomorphic to a closed n-ball B^n. The pair $(p, f|V^{n+k})$ defines an explicit homeomorphism

$$V^{n+k} \xrightarrow[\approx]{(p,f|V^{n+k})} M^k \times D^n.$$

This structure on M^k is called a *framing*, and M is said to be *framed in* $\mathbf{R}^{n+k}$. A choice of basis of the tangent space at $y_0 \in S^n$ pulls back to a set of linearly independent normal vector fields on $M \subset \mathbf{R}^{n+k}$. These will be indicated in Figures 1–3 and 6 below.

Conversely, suppose we have a closed sub-k-manifold $M \subset \mathbf{R}^{n+k}$ with a closed tubular neighborhood V and a homeomorphism h to $M \times D^n$ as above. This is called a *framed sub-k-manifold* of $\mathbf{R}^{n+k}$. Some remarks are in order here.

- The existence of a framing puts some restrictions on the topology of M. All of its charactersitic classes must vanish. In particular it must be orientable.
- A framing can be twisted by a map $g : M \to SO(n)$, where $SO(n)$ denotes the group of orthogonal $n \times n$ matrices with determinant 1. Such matrices act on D^n in an obvious way. The twisted framing

is the composite

$$\begin{array}{ccccc} V & \xrightarrow{h} & M^k \times D^n & \longrightarrow & M^k \times D^n \\ & & (m,x) & \longmapsto & (m, g(m)(x)). \end{array}$$

We will say more about this later.

- If we drop the assumption that M is framed, then the tubular neighborhood V is a (possibly nontrivial) disk bundle over M. The map $M \to y_0$ needs to be replaced by a map to the classifying space for such bundles, $BO(n)$. This leads to unoriented bordism theory, which was analyzed by Thom in [**Tho54**]. Two helpful references for this material are the books by Milnor-Stasheff [**MS74**] and Stong [**Sto68**].

Pontryagin constructs a map $P(M,h) : S^{n+k} \to S^n$ as follows. We regard S^{n+k} as the one point compactification of $\mathbf{R}^{n+k}$ and S^n as the quotient $D^n/\partial D^n$. This leads to a diagram

$$\begin{array}{ccccc} (V, \partial V) & \xrightarrow{h} & M \times (D^n, \partial D^n) & \xrightarrow{p_2} & (D^n, \partial D^n) \\ \big\downarrow & & & & \big\downarrow \\ (\mathbf{R}^{n+k}, \mathbf{R}^{n+k} - \mathrm{int}V) & \longrightarrow & (S^{n+k}, S^{n+k} - \mathrm{int}V) & \xrightarrow{P(M,h)} & (S^n, \{\infty\}) \\ & & S^{n+k} - \mathrm{int}V & \xmapsto{P(M,h)} & \{\infty\} \end{array}$$

The map $P(M,h)$ is the extension of p_2h obtained by sending the compliment of V in S^{n+k} to the point at infinity in S^n. For $n > k$, the choice of the embedding (but not the choice of framing) of M into the Euclidean space is irrelevant. Any two embeddings (with suitably chosen framings) lead to the same map $P(M,h)$ up to continuous deformation.

To proceed further, we need to be more precise about what we mean by continuous deformation. Two maps $f_1, f_2 : X \to Y$ are *homotopic* if there is a continuous map $h : X \times [0,1] \to Y$ (called a *homotopy between f_1 and f_2*) such that

$$h(x,0) = f_1(x) \quad \text{and} \quad h(x,1) = f_2(x).$$

Now suppose $X = S^{n+k}$, $Y = S^n$, and the map h (and hence f_1 and f_2) is smooth with y_0 as a regular value. Then $h^{-1}(y_0)$ is a framed $(k+1)$-manifold N whose boundary is the disjoint union of $M_1 = f^{-1}(y_0)$ and $M_2 = g^{-1}(y_0)$. This N is called a *framed cobordism* between M_1 and M_2, and when it exists the two closed manifolds are said to be *framed cobordant*. An example is shown in Figure 1.

Let $\Omega^{\mathrm{fr}}_{k,n}$ denote the cobordism group of framed k-manifolds in $\mathbf{R}^{n+k}$. The above construction leads to Pontryagin's isomorphism

$$\Omega^{\mathrm{fr}}_{k,n} \xrightarrow{\approx} \pi_{n+k}(S^n).$$

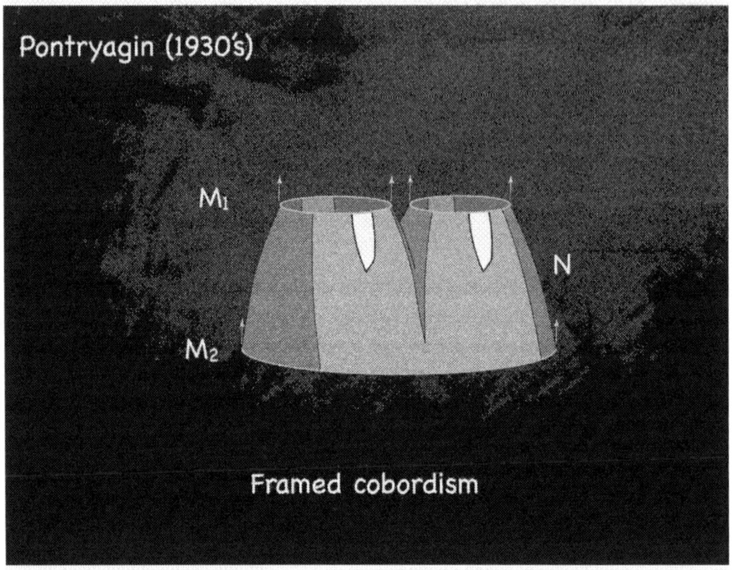

FIGURE 1. A framed cobordism between $M_1 = S^1 \coprod S^1 \subset \mathbf{R}^2$ and $M_2 = S^1 \subset \mathbf{R}^3$ with $N \subset [0,1] \times \mathbf{R}^2$. The normal framings on the circles can be chosen so they extend over N.

First consider the case $k = 0$. Here the 0-dimensional manifold M is a finite set of points in $\mathbf{R}^n$. Each comes with a framing which can be obtained from a standard one by an element in the orthogonal group $O(n)$. We attach a sign to each point corresponding to the sign of the associated determinant. With these signs we can count the points algebraically and get an integer called the *degree of f*. Two framed 0-manifolds are cobordant iff they have the same degree. Figure 2 shows a cobordism between the empty set and a pair of points with opposite signs.

Now consider the case $k = 1$. M is a closed 1-manifold, i.e., a disjoint union of circles. Two framings on a single circle differ by a map from S^1 to the group $SO(n)$, and it is known that

$$\pi_1(SO(n)) = \begin{cases} 0 & \text{for } n = 1 \\ \mathbf{Z} & \text{for } n = 2 \\ \mathbf{Z}/2 & \text{for } n > 2. \end{cases}$$

Figure 2 illustrates the two different framings on S^1 for $n = 2$. It turns about that any disjoint union of framed circles is cobordant to a single framed circle. This can be used to show that

$$\pi_{n+1}(S^n) = \begin{cases} 0 & \text{for } n = 1 \\ \mathbf{Z} & \text{for } n = 2 \\ \mathbf{Z}/2 & \text{for } n > 2. \end{cases}$$

The case $k = 2$ is more subtle. As in the 1-dimensional case we have a complete classification of closed 2-manifolds, and it is only necessary to consider path connected ones. The existence of a framing implies that the surface is orientable, so it is characterized by its genus.

Sidebar 1 The Hopf-Whitehead J-homomorphism

Suppose our framed manifold is S^k with a framing that extends to a D^{k+1}. This will lead to the trivial element in $\pi_{n+k}(S^n)$, but twisting the framing can lead to nontrivial elements. The twist is determined up to homotopy by an element in $\pi_k(SO(n))$. Pontryagin's construction thus leads to the homomorphism

$$\pi_k(SO(n)) \xrightarrow{J} \pi_{n+k}(S^n)$$

introduced by Hopf [**Hop35**] and Whitehead [**Whi42**]. Both source and target known to be independent of n for $n > k+1$. In this case the source group for each k (denoted simply by $\pi_k(SO)$ since n is irrelevant) was determined by Bott [**Bot59**] in his remarkable periodicity theorem. He showed

$$\pi_k(SO) = \begin{cases} \mathbf{Z} & \text{for } k \equiv 3 \text{ or } 7 \bmod 8 \\ \mathbf{Z}/2 & \text{for } k \equiv 0 \text{ or } 1 \bmod 8 \\ 0 & \text{otherwise.} \end{cases}$$

Here is a table showing these groups for $k \leq 10$.

k	1	2	3	4	5	6	7	8	9	10
$\pi_k(SO)$	$\mathbf{Z}/2$	0	$\mathbf{Z}$	0	0	0	$\mathbf{Z}$	$\mathbf{Z}/2$	$\mathbf{Z}/2$	0

In each case where the group is nontrivial, its the image under J of its generator is known to generate a direct summand. In the jth case we denote this image by β_j and its dimension by $\phi(j)$, which is roughly $2j$. The first three of these are the Hopf maps $\eta \in \pi_1^S$, $\nu \in \pi_3^S$ and $\sigma \in \pi_7^S$. After that we have $\beta_4 \in \pi_8^S$, $\beta_5 \in \pi_9^S$, $\beta_6 \in \pi_{11}^S$ and so on.
For the case $\pi_{4m-1}(SO) = \mathbf{Z}$, the image under J is known to be a cyclic group whose order a_m is the denominator of $B_m/4m$, where B_m is the mth Bernoulli number. Details can be found in [**Ada66**] and [**MS74**]. Here is a table showing these values for $m \leq 10$.

m	1	2	3	4	5	6	7	8	9	10
a_m	24	240	504	480	264	65,520	24	16,320	28,728	13,200

If the genus is zero, namely if $M = S^2$, then there is a framing which extends to a 3-dimensional ball. This makes M cobordant to the empty set, which means that the map is *null homotopic* (or, more briefly, *null*), meaning that it is homotopic to a constant map. Any two framings on S^2 differ by an element in $\pi_2(SO(n))$. This group is known to vanish, so any two framings on S^2 are equivalent, and the map $f : S^{n+2} \to S^n$ is null.

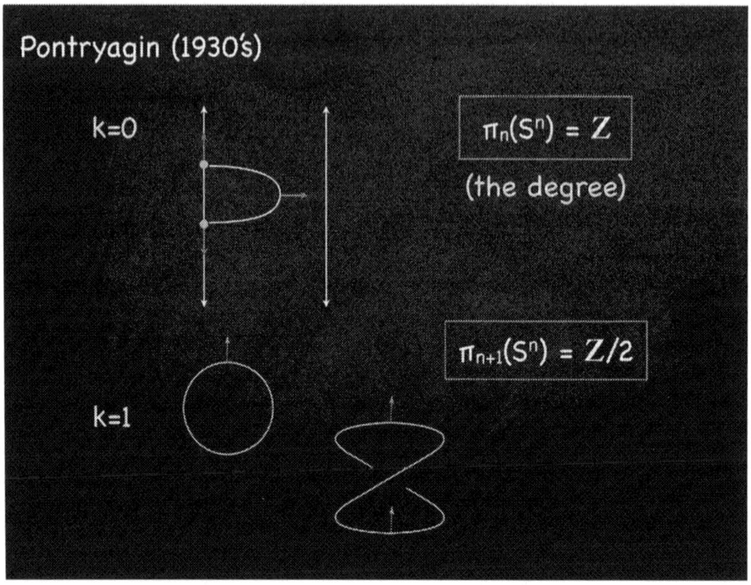

FIGURE 2. The cases $k = 0$ and $k = 1$. The indicated 0-manifold (two points in $\mathbf{R}$ with opposite signs) is framed cobordant (via the yellow line) to the empty set. For $k = 1$, the two circles are framed in $\mathbf{R}^3$. One normal field on each is in the plane of the picture as indicated, and the second (not shown) is pointing out of the plane of the picture toward the reader. Which of these two framings extends to a disk in $\mathbf{R}^3$?

Now suppose the genus is one, as shown in Figure 3. Suppose we can find an embedded arc as shown on which the framing extends to a disk. Then there is a cobordism which effectively cuts along the arc and attaches two disks as shown. This process is called *framed surgery.* If we can do this, then we have converted the torus to a 2-sphere and shown that the map $f : S^{n+2} \to S^n$ is null.

When can we find such a closed curve in M? It must represent a generator of $H_1(M)$ and carry a trivial framing. This leads to a map

$$\varphi : H_1(M; \mathbf{Z}/2) \to \mathbf{Z}/2 \tag{1}$$

defined as follows. Each class in H_1 can be represented by a closed curve which is framed either trivially or nontrivially. It can be shown that homologous curves have the same framing invariant, so φ is well defined. At this point Pontryagin made a famous mistake which went undedected for over a decade: HE ASSUMED THAT φ WAS A HOMOMORPHISM. We now know this is not the case, and we will say more about it below in §1.3. This nonlinearity is illustrated in Figure 4.

On that basis he argued that φ must have a nontrivial kernel, since the source group is $(\mathbf{Z}/2)^2$. Therefore there is a closed curve along which we

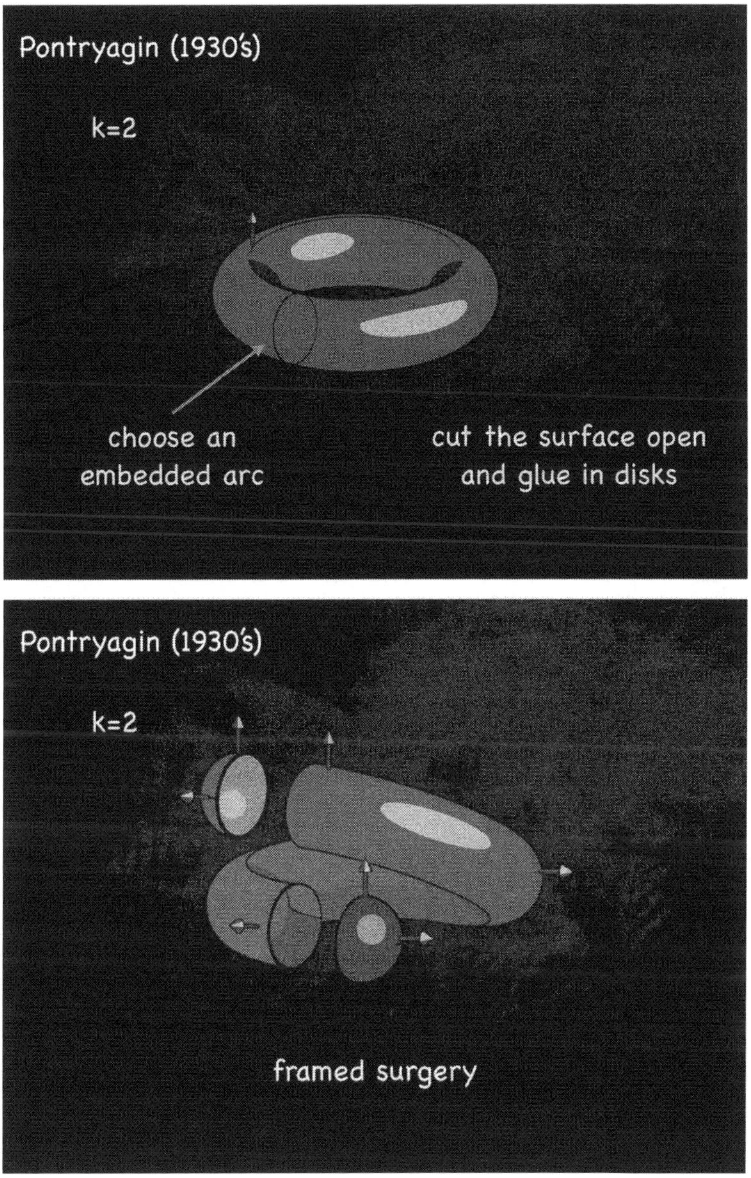

FIGURE 3. The case $k = 2$ and genus 1. If the framing on the embedded arc extends to a disk, then there is a cobordism (called a framed surgery) that converts the torus to a 2-sphere as shown.

can do the surgery shown in Figure 3. It follows that M can be surgered into a 2-sphere, leading to the erroneous conclusion that $\pi_{n+2}(S^n) = 0$ for all n. Freudenthal [**Fre38**] and later George Whitehead [**Whi50**] both proved that it is $\mathbf{Z}/2$ for $n \geq 2$. Pontryagin corrected his mistake in [**Pon50**],

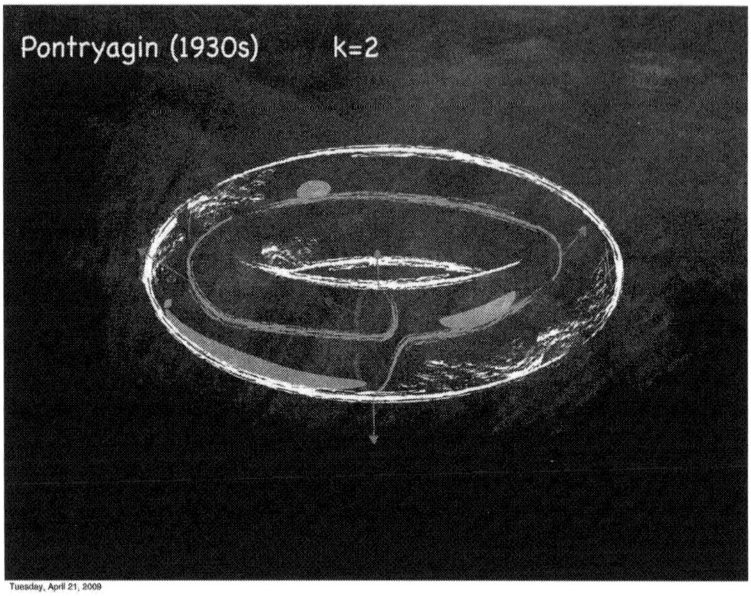

FIGURE 4. The nonlinearity of φ. Even if the framing on the torus is such that its restrictions to the longitudinal and latitudinal circles each extends to a disk, the resulting framing on their sum does not.

and in [**Pon55**] he gave a complete account of the relation between framed cobordism and homotopy groups of spheres.

1.2. Our main result. Our main theorem can be stated in three different but equivalent ways:

- *Manifold formulation:* It says that a certain geometrically defined invariant $\Phi(M)$ (the Arf-Kervaire invariant, to be defined later) on certain manifolds M is always zero.
- *Stable homotopy theoretic formulation:* It says that certain long sought hypothetical maps between high dimensional spheres do not exist.
- *Unstable homotopy theoretic formulation:* It says something about the EHP sequence (to be defined below), which has to do with unstable homotopy groups of spheres.

The problem solved by our theorem is nearly 50 years old. There were several unsuccessful attempts to solve it in the 1970s. They were all aimed at proving the opposite of what we have proved.

Here again is the stable homotopy theoretic formulation.

MAIN THEOREM. *The Arf-Kervaire elements $\theta_j \in \pi^S_{2^{j+1}-2}$ do not exist for $j \geq 7$.*

1.3. The manifold formulation. Let λ be a nonsingular anti-symmetric bilinear form on a free abelian group H of rank $2n$ with mod 2 reduction $\overline{H}$. It is known that $\overline{H}$ has a basis of the form $\{a_i, b_i \colon 1 \leq i \leq n\}$ with

$$\lambda(a_i, a_{i'}) = 0 \quad \lambda(b_j, b_{j'}) = 0 \quad \text{and} \quad \lambda(a_i, b_j) = \delta_{i,j}.$$

In other words, $\overline{H}$ has a basis for which the bilinear form's matrix has the symplectic form

$$\begin{bmatrix} 0 & 1 & & & & & \\ 1 & 0 & & & & & \\ & & 0 & 1 & & & \\ & & 1 & 0 & & & \\ & & & & \ddots & & \\ & & & & & 0 & 1 \\ & & & & & 1 & 0 \end{bmatrix}.$$

A *quadratic refinement of* λ is a map $q : \overline{H} \to \mathbf{Z}/2$ satisfying

$$q(x + y) = q(x) + q(y) + \lambda(x, y)$$

Its *Arf invariant* is

$$\operatorname{Arf}(q) = \sum_{i=1}^{n} q(a_i) q(b_i) \in \mathbf{Z}/2.$$

In 1941 Arf [**Arf41**] proved that this invariant (along with the number n) determines the isomorphism type of q.

An equivalent definition is the "democratic invariant" of Browder. The elements of $\overline{H}$ "vote" for either 0 or 1 by the function q. The winner of the election (which is never a tie) is $\operatorname{Arf}(q)$. Here is a table illustrating this for three possible refinements q, q' and q'' when $\overline{H}$ has rank 2.

x	0	a	b	$a+b$	Arf invariant
$q(x)$	0	0	0	1	0
$q'(x)$	0	1	1	1	1
$q''(x)$	0	1	0	0	0

The value each refinement on $a + b$ is determined by those on a and b, and q'' is isomorphic to q . Thus the vote is three to one in each case. When $\overline{H}$ has rank 4, it is 10 to 6.

Let M be a $2m$-connected smooth closed manifold of dimension $4m + 2$ with a framed embedding in $\mathbf{R}^{4m+2+n}$. We saw above that this leads to a map $f : S^{n+4m+2} \to S^n$ and hence an element in $\pi_{n+4m+2}(S^n)$.

Let $H = H_{2m+1}(M; \mathbf{Z})$, the homology group in the middle dimension. Each $x \in H$ is represented by an immersion $i_x : S^{2m+1} \looparrowright M$ with a stably trivialized normal bundle. H has an antisymmetric bilinear form λ defined in terms of intersection numbers.

In 1960 Kervaire [**Ker60**] defined a quadratic refinement q on its mod 2 reduction in terms of the trivialization of each sphere's normal bundle. The *Kervaire invariant* $\Phi(M)$ is defined to be the Arf invariant of q. In the case $m = 0$, when the dimension of the manifold is 2, Kervaire's q is Pontryagin's map φ of (1).

What can we say about $\Phi(M)$?

- Kervaire [**Ker60**] showed it must vanish when $k = 2$. This enabled him to construct the first example of a topological manifold (of dimension 10) without a smooth structure. This is illustrated in Figure 5. N is a smooth 10-manifold with boundary given as the union of two copies of the tangent disk bundle of S^5. The boundary is homeomorphic to S^9. Thus we can get a closed topological manifold X by gluing on a 10-ball along its common boundary with n, or equivalently collapsing ∂N to a point. X then has nontrivial Kervaire invariant. On the other hand, Kervaire proved that any smooth framed manifold must have trivial Kervaire invariant. Therefore the topological framed manifold X cannot have a smooth structure. Equivalently, the boundary ∂N cannot be diffeomorphic to S^9. It must be an exotic 9-sphere.

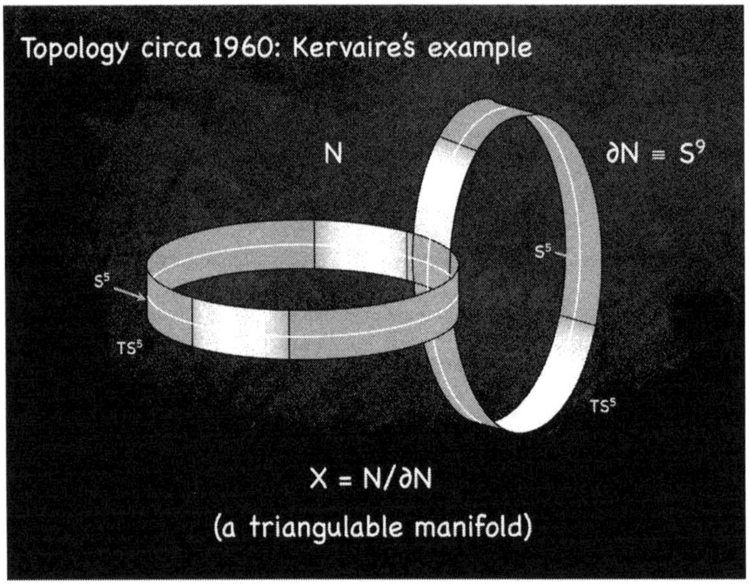

FIGURE 5. Kervaire's example of a nonsmoothable 10-manifold. The manifold N is a smooth 10-manifold with boundary homeomorphic to S^9. The manifold X obtained by collapsing the boundary to a point would have $\Phi(X) = 1$ and is therefore not smooth.

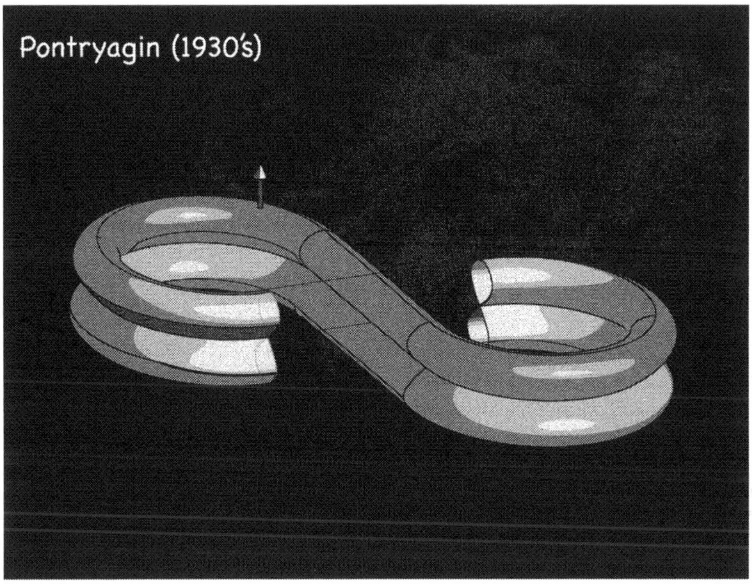

FIGURE 6. A framing on the torus with nontrivial Kervaire invariant. The immersion shown in $\mathbf{R}^3$ is the linear image of an embedding in $\mathbf{R}^4$. This framing on the torus does not exted to any manifold bounded by it.

- For $k = 0$ there is a framing on the torus $S^1 \times S^1 \subset \mathbf{R}^4$ with nontrivial Kervaire invariant. Pontryagin used it in [**Pon50**] (after some false starts in the 30s) to show $\pi_{n+2}(S^n) = \mathbf{Z}/2$ for all $n \geq 2$. It is illustrated in Figure 6. That picture shows a torus immersed in $\mathbf{R}^3$. This immersion is the linear image of an embedding in $\mathbf{R}^4$.
- There are similar constructions for $k = 1$ and $k = 3$, where the framed manifolds are $S^3 \times S^3$ and $S^7 \times S^7$ respectively. Like S^1, S^3 and S^7 are both parallelizable, meaning that their trivial tangent bundles are trivial. The framings can be twisted in such a way as to yield a nontrivial Kervaire invariant.
- Brown-Peterson [**BP66**] showed that it vanishes for all positive even k. This means that apart from the 2-dimensioanl case, any smooth framed manifold with nontrivail Kervaire invariant must a dimension congruent to 6 modulo 8.
- Browder [**Bro69**] showed that it can be nontrivial only if $k = 2^{j-1} - 1$ for some positive integer j. This happens iff the element h_j^2 is a permanent cycle in the Adams spectral sequence, which was originally introduced in [**Ada58**]. (More information about it can be found below in §3.7) in [**Rav86**] and [**Rav04**].) The corresponding element in $\pi^S_{n+2^{j+1}-2}$ is θ_j, the subject of our theorem. *This is the stable homotopy theoretic formulation of the problem.*

- θ_j is known to exist for $1 \le j \le 3$, i.e., in dimensions 2, 6, and 14. In these cases the relevant framed manifold is $S^{2^j-1} \times S^{2^j-1}$ with a twisted framing as discussed above. The framings on S^{2^j-1} represent the elements h_j in the Adams spectral sequence. The Hopf invariant one theorem of Adams [**Ada60**] says that for $j > 3$, h_j is not a permanent cycle in the Adams spectral sequence because it supports a nontrivial differential. (His original proof was not written in this language, but had to do with secondary cohomlogy operations.) This means that for $j > 3$, a smooth framed manifold representing θ_j (i.e., having a nontrivial Kervaire invariant) cannot have the form $S^{2^j-1} \times S^{2^j-1}$.
- θ_j is also known to exist for $j = 4$ and $j = 5$, i.e., in dimensions 30 and 62. In both cases the existence was first established by purely homotopy theoretic means, without constructing a suitable framed manifold. For $j = 4$ this was done by Barratt, Mahowald and Tangora in [**MT67**] and [**BMT70**]. A framed 30-manifold with nontrivial Kervaire invariant was later constructed by Jones [**Jon78**]. For $j=5$ the homotopy theory was done in 1985 by Barratt-Jones-Mahowald in [**BJM84**].
- Our theorem says θ_j does *not* exist for $j \ge 7$. The case $j = 6$ is still open.

Figure 7 illustrates Kervaire's construction of a framed $(4k+2)$-manifold with nontrivial Kervaire invariant. In all cases except $k = 0$, 1 or 3, any framing of this manifold will do because the tangent bundle of S^{2k+1} is nontrivial and leads to a nontrivial invariant. What the picture does not tell us is whether the bounding sphere S^{4k+1} is diffeomorphic to the standard sphere. If it is, then attaching a $(4k+2)$-disk to it will produce a smooth framed manifold with nontrival Kervaire invariant. If it is not, then we have an exotic $(4k+1)$-sphere bounding a framed manifold and hence not detected by framed cobordism.

1.4. The unstable formulation. Assume all spaces in sight are localized and the prime 2. For each $n > 0$ there is a fiber sequence (see §3.1) due to James, [**Jam55**], [**Jam56a**], [**Jam56b**] and [**Jam57**]

$$S^n \xrightarrow{E} \Omega S^{n+1} \xrightarrow{H} \Omega S^{2n+1}. \tag{2}$$

Here $\Omega X = \Omega^1 X$ where $\Omega^k X$ denotes the space of continuous base point preserving maps to X from the k-sphere S^k, known as the kth loop space of X. This leads to a long exact sequence of homotopy groups

$$\cdots \longrightarrow \pi_{m+n}(S^n) \xrightarrow{E} \pi_{m+n+1}(S^{n+1}) \xrightarrow{H} \pi_{m+n+1}(S^{2n+1}) \xrightarrow{P} \pi_{m+n-1}(S^n) \longrightarrow \cdots$$

Here

- E stands for *E*inhängung, the German word for suspension.

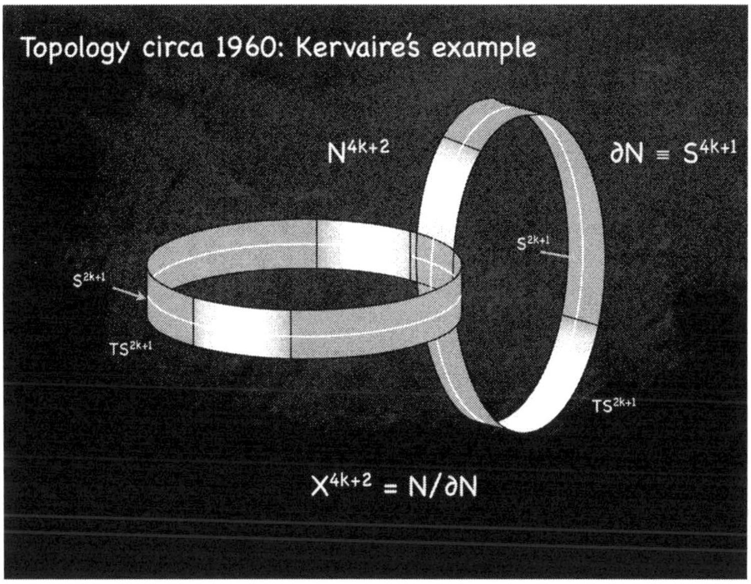

FIGURE 7. Kervaire's example for general k. N is a smooth framed $(4k+2)$-manifold whose boundary is homeomorphic to S^{4k+1}. If ∂N is diffeomorphic to S^{4k+1}, then X is a closed smooth framed $(4k+2)$-manifold with nontrivial Kervaire invariant. We now know this is the case only when $k = 0$, 1, 3, 7, 15 and possibly 31. Otherwise ∂N is an exotic $(4k+1)$-sphere that is a framed boundary.

- H stands for *H*opf invariant.
- P stands for Whitehead *p*roduct.

Assembling these for fixed m and various n leads to a diagram

$$\begin{array}{ccccccc}
 & \pi_{m+n+1}(S^{2n-1}) & & \pi_{m+n+2}(S^{2n+1}) & & \pi_{m+n+3}(S^{2n+3}) & \\
 & \downarrow P & & \downarrow P & & \downarrow P & \\
\cdots \xrightarrow{E} & \pi_{m+n-1}(S^{n-1}) & \xrightarrow{E} & \pi_{m+n}(S^{n}) & \xrightarrow{E} & \pi_{m+n+1}(S^{n+1}) & \xrightarrow{E} \cdots \\
 & \downarrow H & & \downarrow H & & \downarrow H & \\
 & \pi_{m+n-1}(S^{2n-3}) & & \pi_{m+n}(S^{2n-1}) & & \pi_{m+n+1}(S^{2n+1}) &
\end{array}$$

where

- Sequences of arrows labeled H, P, E, H (or any subset thereof) in that order are exact.
- The groups in the top and bottom rows are inductively known, and we can compute those in the middle row by induction on n.
- The groups in the top and bottom rows vanish for large n, making E an isomorphism.
- An element in the middle row has trivial suspension (is killed by E) iff it is in the image of P.

- It desusupends (is in the image of E) iff its Hopf invariant (image under H) is trivial.

When $m = n - 1$ this diagram is

$$\begin{array}{ccccccccc}
 & & & & \pi_{2n+1}(S^{n+1}) & & & & \\
 & & & & \downarrow H & & & & \\
 & & \pi_{2n}(S^{2n-1}) & & \mathbf{Z} & & 0 & & \\
 & & \downarrow P & & \downarrow P & & \downarrow P & & \\
\cdots & \xrightarrow{E} & \pi_{2n-2}(S^{n-1}) & \xrightarrow{E} & \pi_{2n-1}(S^{n}) & \xrightarrow{E} & \pi_{2n}(S^{n+1}) & \xrightarrow{E} & \cdots \\
 & & \downarrow H & & \downarrow H & & \downarrow H & & \\
 & & \pi_{2n-2}(S^{2n-3}) & & \mathbf{Z} & & 0 & &
\end{array}$$

The image under P of the generator of the upper $\mathbf{Z}$ is denoted by $w_n \in \pi_{2n-1}(S^n)$ and is called the *Whitehead square.*

- When n is even, $H(w_n) = 2$ and w_n has infinite order.
- w_n is trivial for $n = 1$, 3 and 7. In those cases the generator of the upper $\mathbf{Z}$ is the Hopf invariant (image under H) of one of the three Hopf maps in $\pi_{2n+1}(S^{n+1})$,

$$S^3 \xrightarrow{\eta} S^2, \qquad S^7 \xrightarrow{\nu} S^4 \quad \text{and} \quad S^{15} \xrightarrow{\sigma} S^8.$$

- For other odd values of n, twice the generator of the upper $\mathbf{Z}$ is $H(w_{n+1})$, so w_n has order 2.
- It turns out that w_n is divisible by 2 iff $n = 2^{j+1} - 1$ and θ_j exists, in which case $w_n = 2\theta_j$.
- Each Whitehead square $w_{2n+1} \in \pi_{4n+1}(S^{2n+1})$ (except the cases $n = 0$, 1 and 3) desuspends to a lower sphere until we get an element with a nontrivial Hopf invariant, which is always some β_j (see Sidebar 1). More precisely we have

$$H(w_{(2s+1)2^j-1}) = \beta_j$$

for each $j > 0$ and $s \geq 0$. This result is essentially Adams' 1962 solution to the vector field problem [**Ada62**].

Recall the EHP sequence

$$\cdots \longrightarrow \pi_{m+n}(S^n) \xrightarrow{E} \pi_{m+n+1}(S^{n+1}) \xrightarrow{H} \pi_{m+n+1}(S^{2n+1}) \xrightarrow{P} \pi_{m+n-1}(S^n) \longrightarrow \cdots$$

Given some $\beta_j \in \pi_{\phi(j)+2n+1}(S^{2n+1})$ for $\phi(j) < 2n$, one can ask about the Hopf invariant of its image under P, which vanishes when β_j is in the image of H. In most cases the answer is known and is due to Mahowald, [**Mah67**] and [**Mah82**]. The remaining cases have to do with θ_j. The answer that he had hoped for is the following, which can be found in [**Mah67**]. (To our knowledge, Mahowald never referred to this as the World Without End Hypothesis. We chose that term to emphasize its contrast with the Doomsday Hypothesis.)

World Without End Hypothesis (Mahowald 1967).

- *The Arf-Kervaire element* $\theta_j \in \pi^S_{2^{j+1}-2}$ *exists for all* $j > 0$.
- *It desuspends to* $S^{2^{j+1}-1-\phi(j)}$ *and its Hopf invariant is* β_j.
- *Let* $j, s > 0$ *and suppose that* $m = 2^{j+2}(s+1) - 4 - \phi(j)$ *and* $n = 2^{j+1}(s+1) - 2 - \phi(j)$. *Then* $P(\beta_j)$ *has Hopf invariant* θ_j.

This describes the systematic behavior in the EHP sequence of elements related to the image of J, and the θ_j are an essential part of the picture. Because of our theorem, *we now know that this hypothesis is incorrect.*

1.5. Questions raised by our theorem. *EHP sequence formulation.* The World Without End Hypothesis was the nicest possible statement of its kind given all that was known prior to our theorem. Now we know it cannot be true since θ_j does not exist for $j \geq 7$. *This means the behavior of the indicated elements* $P(\beta_j)$ *for* $j \geq 7$ *is a mystery.*

Adams spectral sequence formulation. (See §3.7.) We now know that the h_j^2 for $j \geq 7$ are not permanent cycles, so they have to support nontrivial differentials. *We have no idea what their targets are.*

Manifold formulation. Here our result does *not* lead to any obvious new questions. It appears rather to be the final page in the story.

Our method of proof offers a new tool for studying the stable homotopy groups of spheres. We look forward to learning more with it in the future.

2. Our strategy

2.1. Ingredients of the proof. Our proof has several ingredients.

- It uses methods of stable homotopy theory, which means it uses spectra instead of topological spaces. For more information about this see §4. Recall that a space X has a homotopy group $\pi_k(X)$ for each positive integer k. A spectrum X has an abelian homotopy group $\pi_k(X)$ *defined for every integer* k. For the sphere spectrum S^0, $\pi_k(S^0)$ is the stable k-stem homotopy group π_k^S. The hypothetical θ_j is an element of this group for $k = 2^{j+1} - 2$.
- It uses complex cobordism theory. This is a branch of algebraic topology having deep connections with algebraic geometry and number theory. It includes some highly developed computational techniques that began with work by Milnor [**Mil60**], Novikov ([**Nov60**], [**Nov62**] and [**Nov67**]) and Quillen [**Qui69**] in the 60s. A pivotal tool in the subject is the theory of formal group laws. On this subject the definitive reference is Hazewinkel's book [**Haz78**]. A much briefer account covering the most relevant aspects of the subject can be found in [**Rav86**, Appendix 2].

- It also makes use of newer less familiar methods from equivariant stable homotopy theory. A helpful introduction to this subject is the paper of Greenlees-May [**GM95**]. This means there is a finite group G (a cyclic 2-group) acting on all spaces in sight, and all maps are required to commute with these actions. When we pass to spectra, we get homotopy groups indexed not just by the integers $\mathbf{Z}$, *but by* $RO(G)$, *the orthogonal representation ring of* G. Our calculations make use of this richer structure.

2.2. The spectrum Ω. We will produce a map $S^0 \to \Omega$, where Ω is a nonconnective spectrum (meaning that it has nontrivial homotopy groups in arbitrarily large negative dimensions) with the following properties.

(i) *Detection Theorem.* It has an Adams-Novikov spectral sequence (which is a device for calculating homotopy groups) in which the image of each θ_j is nontrivial. *This means that if* θ_j *exists, we will see its image in* $\pi_*(\Omega)$.

(ii) *Periodicity Theorem.* It is 256-periodic, meaning that $\pi_k(\Omega)$ depends only on the reduction of k modulo 256.

(iii) *Gap Theorem.* $\pi_k(\Omega) = 0$ for $-4 < k < 0$. This property is our zinger. Its proof involves a new tool we call the slice spectral sequence.

If $\theta_7 \in \pi_{254}(S^0)$ exists, (i) implies it has a nontrivial image in $\pi_{254}(\Omega)$. On the other hand, (ii) and (iii) imply that $\pi_{254}(\Omega) = 0$, so θ_7 cannot exist. The argument for θ_j for larger j is similar, since $|\theta_j| = 2^{j+1} - 2 \equiv -2 \bmod 256$ for $j \geq 7$.

2.3. How we construct Ω. Our spectrum Ω will be the fixed point spectrum for the action of C_8 (the cyclic group of order 8) on an equivariant spectrum $\tilde{\Omega}$.

To construct it we start with the complex cobordism spectrum MU. It can be thought of as the set of complex points of an algebraic variety defined over the real numbers. This means that it has an action of C_2 defined by complex conjugation. The fixed point set of this action is the set of real points, known to topologists as MO, the unoriented cobordism spectrum. In this notation, U and O stand for the unitary and orthogonal groups.

To get a C_8-spectrum, we use the following general construction for getting from a space or spectrum X acted on by a group H to one acted on by a larger group G containing H as a subgroup. Let

$$Y = \mathrm{Map}_H(G, X),$$

the space (or spectrum) of H-equivariant maps from G to X. Here the action of H on G is by right multiplication, and the resulting object has an action of G by left multiplication. As a set, $Y = X^{|G/H|}$, the $|G/H|$-fold Cartesian power of X. A general element of G permutes these factors, each of which is left invariant by the subgroup H.

In particular we get a C_8-spectrum

$$MU^{(4)} = \mathrm{Map}_{C_2}(C_8, MU).$$

This spectrum is not periodic, but it has a close relative $\tilde{\Omega}$ which is.

3. Some classical algebraic topology

3.1. Fibrations. A map $p : E \to B$ is a *fibration* (sometimes called a Hurewicz fibration [**Hur35**][**Hur36**]) if the following commutuative diagram can always be completed:

(3)
$$\begin{array}{ccccc} x & \quad & X & \xrightarrow{\tilde{f}_0} & E \\ \downarrow & & \downarrow{\scriptstyle i} & \overset{\tilde{f}}{\nearrow} & \downarrow{\scriptstyle p} \\ (x,0) & & X \times I & \xrightarrow{f} & B, \end{array}$$

where I denotes the closed unit interval $[0,1]$. In other words, given maps f and $\tilde{f}_0$ as shown, one can always find a map $\tilde{f}$ such that $\tilde{f}i = \tilde{f}_0$ and $p\tilde{f} = f$. This is called the *homotopy lifting property* because it says that if one end of the homotopy f can be lifted from B to E via $\tilde{f}_0$, then we can find a lifting$\tilde{f}$ of the entire homotopy.

Experience has shown that it often suffices to assume (3) holds only in the case where X is an n-dimensional disk D^n for any n. In this case p is called a *Serre fibration* [**Ser51**]. It is enough to establish the long exact sequence of homotopy groups (4).

One example of a fibration is a *fiber bundle* [**Ste99**], a map for which

(i) the preimage $p^{-1}(b)$ for each $b \in B$ is homeomorphic to a space F called the *fiber of* p and
(ii) each $b \in B$ has a neighborhood U such that $p^{-1}(U)$ is homeomorphic to $U \times F$ in such a way that the composition of p with the homeomorphism is projection onto U.

In this case there is a long exact sequence of homotopy groups

(4)
$$\cdots \longrightarrow \pi_n(F) \xrightarrow{i_*} \pi_n(E) \xrightarrow{p_*} \pi_n(B) \xrightarrow{\partial} \pi_{n-1}(F) \longrightarrow \cdots$$

where $i : F \to E$ is the inclusion of the fiber and ∂ is a certain boundary homomorphism. More details can be found in any textbook on algebraic topology, such as Hatcher [**Hat02**], May [**May99**] or Gray [**Gra75**].

There is a way to replace any map $p : E \to B$ by a fibration between homotopy equivalent spaces and construct a homotopy theoretic fiber. First we replace B by the *mapping cylinder* $\tilde{B} = M_p$, which is the quotient of the disjoint union

$$(E \times I) \bigcup B$$

obtained by identifying $(e,1) \in E \times I$ with $p(e) \in B$. For technical reasons it is convenient to assume that E and B have base points e_0 and b_0 with

$p(e_0) = b_0$, and to collapse $\{e_0\} \times I$ to a point which becomes the base point in $\tilde{B}$. The inclusion of B into $\tilde{B}$ is a homotopy equivalence.

Next we replace E by the space $\tilde{E}$ of paths (continuous maps from the unit interval I) in $\tilde{B}$ that begin in $E \times \{0\}$. The inclusion of E into $\tilde{E}$ via paths of the form $t \mapsto (e, t)$ is again a homotopy equivalence. Finally we define $\tilde{p} : \tilde{E} \to \tilde{B}$ by sending a path to its endpoint. Then the diagram

$$\begin{array}{ccc} E & \xrightarrow{\simeq} & \tilde{E} \\ \downarrow{\scriptstyle p} & & \downarrow{\scriptstyle \tilde{p}} \\ B & \xrightarrow{\simeq} & \tilde{B} \end{array}$$

commutes, the horizontal maps are homotopy equivalences, and $\tilde{p}$ is a fibration.

Now let $\tilde{F} \subset \tilde{E}$ be the space of paths starting in $E \times \{0\}$ and ending at the base point; we denote the inclusion map by $\tilde{i}$. This is the *homotopy theoretic fiber of* p and there is a long exact sequence similar to (4),

$$\cdots \longrightarrow \pi_n(\tilde{F}) \xrightarrow{\tilde{i}_*} \pi_n(\tilde{E}) \xrightarrow{\tilde{p}_*} \pi_n(\tilde{B}) \xrightarrow{\tilde{\partial}} \pi_{n-1}(\tilde{F}) \longrightarrow \cdots$$

A *fiber sequence* (such as (2)) is any composite

$$W \xrightarrow{f} X \xrightarrow{g} Y$$

where f is equivalent to the inclusion of the homotopy theoretic fiber of g.

When the map p is inclusion of the base point b_0 of B, we find that

- $\tilde{B} = B$,
- $\tilde{E} = PB$, the space of paths in B starting at b_0, which is contractible, and
- $\tilde{F} = \Omega B$, the space of closed path starting and ending at b_0, which is called the *loop space of* B.

This is called the *path fibration of* B. The long exact sequence in this case gives an isomorphism

$$\pi_n(\Omega B) \approx \pi_{n+1}(B).$$

When we apply this constuction to the map $\tilde{d}$, we find that its homotopy theoretic fiber is equivalent to ΩB. The homotopy theoretic fiber of the map $\Omega B \to \tilde{F}$ is equivalent to ΩE, and so on.

3.2. Cofibrations. A *cofibration* is a map $i : A \to X$ satisfying a diagram dual to (3), namely

(5)
$$\begin{array}{cccc} \omega(0) & Y & \xleftarrow{\tilde{f}_0} & X \\ \uparrow & \uparrow{\scriptstyle p_0} & \swarrow{\scriptstyle \tilde{f}} & \uparrow{\scriptstyle i} \\ \omega & Y^I & \xleftarrow{f} & A \end{array}$$

Sidebar 2 CW complexes

A *CW-complex* (first introduced by Henry Whitehead in [**Whi49**]) X is a space constructed as a union of skeleta X^n defined as follows. X^0 is a discrete set. X^n is obtained from X^{n-1} by attaching a collection of n-disks $\{D^n_\alpha\}$ by identifying points on their boundaries with points in X^{n-1} using maps $f_\alpha : \partial D^n_\alpha \to X^{n-1}$ called *attaching maps.* X^n is the cofiber (see §3.2) of the map to X^{n-1} from the disjoint union of the $(n-1)$-spheres ∂D^n_α. The n-disks are called *n-cells.* A CW-complex is said to be *finite* if its has only a finite number of cells altogether, and to have *finite type* if it has only finitely many in each dimension.

This collection of spaces is general enough to include all manifolds and all real and complex algebraic varieties. Indeed most homotopy theorists never have to deal with spaces that are not at least homotopy equivalent to CW-complexes. They have the following convenient properties:

- The product of two CW-complexes is a CW-complex.
- A map $f : X \to Y$ of CW-complexes is a homotopy equivalence (which means there is a map $g : Y \to X$ such that fg and gf are homotopic to the identities maps on Y and X) iff it is a weak homotopy equivalence, meaning that it induces an isomorphism in all homotopy groups.
- Any such map is homotopic to one that sends X^n to Y^n for all n.
- The space of maps $X \to Y$, while not a CW-complex itself in general, always has the homotopy type of one by a theorem of Milnor [**Mil59**]. The same holds if one requires that certain subcomplexes of X map to certain subcomplexes of Y. For example, ΩS^{n+1} is equivalent to a CW-complex with a single cell in each dimension divisible by n, and no others.
- $H_*(X)$ can be computed in terms of a *cellular chain complex* $C_*(X)$ in which the nth chain group $C_n(X)$ is free abelian on the n-cells of X.

Here Y^I denotes the space of maps $\omega : I \to Y$, i.e., all paths in Y. The map $f : A \to Y^I$ is formally equivalent to a homotopy $A \times I \to Y$. $\tilde{f}_0$ is an extension of one end of it to X and the hypothetical $\tilde{f}$ is an extension of all of it to X, so this is called the *homotopy extension property.*

The inclusion i of a subcomplex A into a CW-complex X (see Sidebar 2) is always a cofibration. It has a *cofiber* C_i (also know as the *mapping cone*), which, like the mapping cylinder above, is the quotient of the disjoint union

$$(A \times I) \bigcup X$$

obtained by identifying $(a,1) \in A \times I$ with $i(a) \in X$ (as in the mapping cylinder) and by collapsing all of $A \times \{0\}$ to a point. As before, if a base point is desired, it can be obtained by collapsing the path along the base point of A to a point. We denote the inclusion map $X \to C_i$ by j .

Then the reduced homology $\overline{H}_*(C_i)$ of the cofiber is the same as the relative homology $H_*(X,A)$. There is a long exact sequence

(6)

$$\cdots \longrightarrow H_n(A) \xrightarrow{i_*} H_n(X) \xrightarrow{j_*} H_n(C_i) \xrightarrow{\partial} H_{n-1}(A) \longrightarrow \cdots$$

The procedure for replacing an arbitrary map $i : A \to X$ by a homotopy equivalent cofibration is easier here than it was for fibrations above. All we have to do is replace X by the mapping cylinder $\tilde{X} = M_i$. This makes the evident inclusion $\tilde{i} : A \to \tilde{X}$ a cofibration, and the mapping cone C_i is its cofiber. The simplicity of this construction renders the term "homotopy theoretic cofiber" unnecessary. C_i is called the simply the *cofiber of the map* i. The cofiber of the map $j : M_i \to C_i$ is easily seen to be the suspension ΣA.

A *cofiber sequence* is any composite

$$W \xrightarrow{f} X \xrightarrow{g} Y$$

where g is equivalent to the map j to C_f. One has similar notions in the category of spectra, to be defined in §4. In that world fiber sequence and cofiber sequences are the same: W is equivalent to the fiber of g iff Y is equivalent to the cofiber of f. This means we have long exact sequences in both homotopy and homology.

3.3. Eilenberg-Mac Lane spaces and cohomology operations.

THEOREM 1. **Eilenberg-Mac Lane spaces and cohomology.** *For any CW-complex X, positive integer n and discrete abelian group A, there is a natural isomorphism between the cohomology group $H^n(X;A)$ and the group of homotopy classes of maps $[X, K(A,n)]$,*

$$H^n(X;A) \approx [X, K(A,n)] \tag{7}$$

where $K(A,n)$ is the Eilenberg-Mac Lane space constructed in Sidebar 3.

The group structure on the set of homotopy classes arises from a map $K(A,n) \times K(A,n) \to K(A,n)$ with suitable properties. This isomorphism holds if X is also an Eilenberg-Mac Lane space. When it is $K(A,n)$, we get the identity map on the right corresponds to a canonical element ι_n on the left called the *fundamental class.*

Let $\theta \in H^{n+k}(K(A,n);A')$ and $x \in H^n(X;A)$ for $k \geq 0$. (The case $k < 0$ is uninteresting because the cohomology group is trivial.) Using (7), these correspond to maps

$$X \xrightarrow{x} K(A,n) \xrightarrow{\theta} K(A', n+k),$$

so the composite θx corresponds to an element in $H^{n+k}(X; A')$. Hence θ gives us a natural transformation from the functor $H^n(\cdot; A)$ to $H^{n+k}(\cdot; A')$ called a *cohomology operation.* It may or may not be a group homomorphism. When it is, we say it is *additive.*

Now assume that all cohomology groups have coefficients in $\mathbf{Z}/2$ and let $K_n = K(\mathbf{Z}/2, n)$. Let $s_n : \Sigma K_n \to K_{n+1}$ be adjoint to the equivalence $K_n \to \Omega K_{n+1}$. It induces a homomorphism

$$H^{n+k+1}(K_{n+1}) \xrightarrow{s_n^*} H^{n+k+1}(\Sigma K_n) = H^{n+k}(K_n) \ni \theta$$

It is known that θ is additive iff it is in the image of s_n^*. In that case it is also in the image of a similar map from $H^{n+k+t}(K_{n+t})$ for any $t > 0$, and there is a way to choose these preimages canonically. Such an operation θ is said to be *stable.*

3.4. The Steenrod algebra. Here is an important example of a mod 2 stabel operation. Let $\theta = \iota_n^2 \in H^{2n}(K_n)$. The corresponding cohomology operation sends x to x^2, which is additive since we are working mod 2. Since it is additive, it is stable and we have similar operations

$$H^{n+t}(X) \xrightarrow{Sq^n} H^{2n+t}(X)$$

for all $t \geq 0$. This is called the *nth Steenrod squaring operation* [**Ste62**]. These operations have been studied extensively and it is known that any mod 2 stable operation can be expressed in terms of them. They have the following properties:

THEOREM 2. **Properties of mod 2 Steenrod operations**

(i) Sq^0 *is the identity map.*

(ii) CARTAN FORMULA.

$$Sq^n(xy) = \sum_{0 \leq i \leq n} Sq^i(x) Sq^{n-i}(y).$$

(iii) ADEM RELATION. *For* $0 < a < 2b$,

$$Sq^a Sq^b = \prod_{j=0}^{j=[a/2]} \binom{b-i-2j}{a-2j} Sq^{a+b-j} Sq^j.$$

(iv) UNSTABLE CONDITION. $Sq^n(x) = x^2$ *for* $x \in H^n$ *and* $Sq^n(x) = 0$ *for* $x \in H^i$ *with* $i < n$.

For $n > 0$, the Adem relation gives $Sq^{2n+1} = Sq^1 Sq^{2n}$ and

$$Sq^{(2n+1)2^{k+1}} = Sq^{2^{k+1}} Sq^{2^{k+2}n} + Sq^{2^{k+2}n+2^k} Sq^{2^k} \quad \text{for } k \geq 0,$$

so Sq^i is decomposable unless i is a power of 2.

A monomial $Sq^I = Sq^{i_1} Sq^{i_2} \cdots Sq^{i_m}$ is *admissible* if

$$i_1 \geq 2i_2,\ i_2 \geq 2i_3, \ldots,\ i_{m-1} \geq 2i_m. \tag{8}$$

Repeated use of the Adem relation will convert any monomial to a sum of admissible ones.

The *mod 2 Steenrod algebra* $\mathcal{A}$ is the algebra of all mod 2 stable cohomology operations. It is the associative algebra generated by the Sq^n for $n > 0$ and subject to the Adem relation. The admissible monomials form a basis for it (the *Adem basis*) and it is generated as an algebra by the elements Sq^{2^j} for $j \geq 0$.

For any space X, $H^*(X)$ is an $\mathcal{A}$-module subject to the Cartan formula and the unstable condition. For a spectrum X, these restrictions are vacuous, so we just get an $\mathcal{A}$-module.

We will now describe $H^*(K_n)$ for all n. The *excess* of an admissible monomial Sq^I as in (8) is

$$e(I) = (i_1 - 2i_2) + (i_2 - 2i_3) + \cdots + (i_{m-1} - 2i_m) = i_1 - i_2 - i_3 - \cdots - i_m.$$

THEOREM 3. **Mod 2 cohomology of mod 2 Eilenberg-Mac Lane spaces.**

$$H^*(K_n) = \mathbf{Z}/2[Sq^I\iota_n\colon e(I) < n,\ I\ \textit{admissible}].$$

The only admissible monomial with excess 0 is 1, so the theorem says

$$H^*(K_1) = \mathbf{Z}/2[\iota_1],$$

which is consistent with the fact that $K_1 = \mathbf{R}P^\infty$.

The set of admissible monomials with excess 1 is

$$\left\{Sq^1,\ Sq^2Sq^1,\ Sq^4Sq^2Sq^1,\ Sq^8Sq^4Sq^2Sq^1,\ \dots\right\},$$

which leads to

$$H^*(K_2) = \mathbf{Z}/2[x_2,\ x_3,\ \dots, x_{1+2^k}, \dots]$$

where $x_2 = \iota_2$ and $x_{1+2^{k+1}} = Sq^{2^k}x_{1+2^k}$.

This entire discussion has an odd primary analog, but we do not need it here.

3.5. Milnor's formulation. The results of this section are due to Milnor [**Mil58**].

The Cartan formula (Theorem 2 (ii)) leads to a coproduct

$$\begin{array}{ccc} \mathcal{A} & \longrightarrow & \mathcal{A}\otimes\mathcal{A} \\ Sq^n & \longmapsto & \sum_{0\leq i\leq n} Sq^i\otimes Sq^{n-i} \end{array}$$

which is an algebra map and is cocommutative. This leads to a commutative algebra structure on the dual

$$\mathcal{A}_* = \mathrm{Hom}_{\mathbf{Z}/2}(\mathcal{A}, \mathbf{Z}/2).$$

The noncommutative product on $\mathcal{A}$, which is a map $\mathcal{A}\otimes\mathcal{A}\to\mathcal{A}$, translates to a noncommutative coproduct on $\mathcal{A}_*$ which we denote by Δ.

THEOREM 4. **The structure of the dual Steenrod algebra.** *As an algebra,*

$$\mathcal{A}_* = \mathbf{Z}/2[\xi_1, \xi_2, \ldots]$$

where the dimension of ξ_j *is* $2^j - 1$. *The coproduct is*

$$\Delta(\xi_n) = \sum_{0 \le i \le n} \xi_{n-i}^{2^i} \otimes \xi_i \quad \textit{where } \xi_0 = 1.$$

This coproduct formula is equivalent to (and much easier to remember than) the Adem relation, but proving this explicitly is difficult.

3.6. Serre's method of computing homotopy groups. Let X be a $(n-1)$-connected space for $n > 1$ with $\pi_n(X)$, nontrivial, and let $L_0 = K(\pi_n(X), n)$. The Hurewicz theorem [**Hur35**] [**Hur36**] says that $\pi_n(X)$ is isomorphic to $H_n(X; \mathbf{Z})$, and Theorem 1 then implies there is a map $f_0 : X \to L_0$ inducing an isomorphism in π_n. Let X_1 be its fiber as explained in 3.1. The the long exact sequence of homotopy groups (4) implies that

$$\pi_i(X_1) = \begin{cases} 0 & \text{for } i \le n \\ \pi_i(X) & \text{otherwise} \end{cases}$$

The Serre spectral sequence of [**Ser51**] is a device for computing the homology of one of the spaces in a fiber sequence in terms of the other two. In theory we can use it to find the homology of X_1 and hence its first nontrivial homotopy group, which lies somewhere above dimension n. Then we can treat X_1 the same way we treated X and find a map $f_1 : X_1 \to L_1$ where L_1 is the appropriate Eilenberg-Mac Lane space. Then we could use the Serre spectral sequence to find the homology of the fiber X_2, and so on. Continuing in this manner we could get a diagram

$$(9) \qquad \begin{array}{ccccccccc} X & \xleftarrow{\;g_0\;} & X_1 & \xleftarrow{\;g_1\;} & X_2 & \xleftarrow{\;g_2\;} & X_3 & \xleftarrow{\;g_3\;} & \cdots \\ \downarrow{\scriptstyle f_0} & & \downarrow{\scriptstyle f_1} & & \downarrow{\scriptstyle f_2} & & \downarrow{\scriptstyle f_3} & & \\ L_0 & & L_1 & & L_2 & & L_3 & & \end{array}$$

where

(i) each L_i is an Eilenberg-Mac Lane space,
(ii) f_i induces an isomorphism in the first nontrivial homotopy group of X_i and
(ii) X_{i+1} is the fiber of f_i.

This method requires knowing the homology of all spaces in sight, at least through the range of dimensions in which one hopes to compute. It was used with brilliant effect by Serre in [**Ser51**] to calculate many previously unknown homotopy groups of spheres, but a few years later Adams found a better method [**Ada58**] for doing this.

Before turning to it, we need to make one observation about the Serre spectral sequence. In each of the fiber sequences in (9), below dimension $2n$ it simplifies to a long exact sequence

(10)
$$\cdots \longrightarrow H_i(X_{s+1}) \xrightarrow{(g_{s+1})_*} H_i(X_s) \xrightarrow{f_{s*}} H_i(L_s) \xrightarrow{\partial} H_{i-1}(X_{s+1}) \longrightarrow \cdots$$

3.7. The Adams spectral sequence. In this section we return to our convention that all cohomology groups are reduced and have coefficients in $\mathbf{Z}/2$.

We modify Serre's diagram (9) below dimension $2n$ in the following ways:

(i) Each L_s is a product of mod 2 Eilenberg-Mac Lane spaces K_m, with possibly different values of m (all at least n) for the various factors. Theorem 3 implies that in our range $H^*(K_m) = \Sigma^m \mathcal{A}$, the mth suspension of the Steenrod algebra.

(ii) L_s and f_s are chosen so that the induced map in cohomology is onto. This can always be done, but not uniquely. Choose a set of elements in $H^*(X_s)$ which generate it as an $\mathcal{A}$-module. Each one corresponds to a map to some K_m, so collectively they give a map to the product of K_ms, which is L_s.

A diagram of the form (9) having these two properties is called a *mod 2 Adams resolution for* X. When X is a CW-complex of finite type, it leads to a method for computing $\pi_*(X)$ modulo odd primary torsion below dimension $2n$.

The requirement on $H^*(f_s)$ means that $H^*(g_{s+1})$ is trivial, and the cohomological analog of (10) reduces to a short exact sequence of $\mathcal{A}$-modules

$$0 \longrightarrow H^*(\Sigma X_{s+1}) \longrightarrow H^*(L_s) \xrightarrow{f_s^*} H^*(X_s) \longrightarrow 0$$

where $H^*(L_s)$ is free. These can be spliced together for various s, giving us a long exact sequence

(11)
$$0 \longleftarrow H^*(X) \longleftarrow H^*(L_0) \longleftarrow H^*(\Sigma L_1) \longleftarrow H^*(\Sigma^2 L_2) \longleftarrow \cdots$$

This is a free $\mathcal{A}$-resolution of $H^*(X)$. We can recover $\pi_*(L_s)$ from its cohomology by applying the functor $\mathrm{Hom}_{\mathcal{A}}(\cdot, \mathbf{Z}/2)$. Doing this to (11) gives us a cochain complex

(12)
$$\pi_*(L_0) \longrightarrow \pi_*(\Sigma L_1) \longrightarrow \pi_*(\Sigma^2 L_2) \longrightarrow \cdots$$

THEOREM 5. **The Adams spectral sequence.** *For an $(n-1)$-connected space X, in dimensons less than $2n$ the cohomology of (12) is independent of the choices of the L_s, and is by definition* $\mathrm{Ext}_{\mathcal{A}}(H^*(X), \mathbf{Z}/2)$. *This is the*

E_2-term of the Adams spectral sequence, which converges to $\pi_*(X)$ modulo odd torsion. More specifically,

$$E_2^{s,t} = \operatorname{Ext}_{\mathcal{A}}^{s,t}\left(H^*(X), \mathbf{Z}/2\right),$$

$d_r : E_r^{s,t} \to E_r^{s+r,t+r-1}$, and $E_\infty^{s,t}$ is a subquotient of $\pi_{t-s}(X)$.

This theorem is crying out to be reformulated in term of spectra (to be defined in the next section), which is how it is usually stated. Then the caveats about n can be eliminated, provided X is connective and of finite type.

The example of interest to us is the sphere spectrum, for which $E_2^{*,*} = \operatorname{Ext}_{\mathcal{A}}^{*,*}(\mathbf{Z}/2, \mathbf{Z}/2)$. This group is difficult to compute but has been studied extensively, and we will abbreviate it simply by $\operatorname{Ext}^{*,*}$. Here is the first stage of the free resolution of (11) in this case.

$$0 \longleftarrow \mathbf{Z}/2 \xleftarrow{\ \epsilon\ } \mathcal{A} \longleftarrow \mathcal{A}\{x_j : j \geq 0\} \longleftarrow \cdots$$

$$Sq^{2^j} \longleftarrow\!\!\shortmid\; x_j$$

We know that $\ker \epsilon$ is generated as an $\mathcal{A}$-module by $\{x_j : j \geq 0\}$ because these elements generate $\mathcal{A}$ as an algebra.

This leads to elements $h_j \in \operatorname{Ext}^{1,2^j}$, and they are known to form a basis of Ext^1. The group Ext has a ring structure, and the following set forms a basis of Ext^2:

$$\{h_j h_k : 0 \leq j \leq k,\ k \neq j+1\}.$$

The following facts are known about these elements.

- The h_j for $0 \leq j \leq 3$ are nontrivial permanent cycles, meaning that they detects elements in π_*^S. h_0 detects the degree 2 map in π_0^S, and the other three detect maps constructed by Hopf in [**Hop30**] and [**Hop35**].
- Adams [**Ada60**] showed that the h_j for $j > 3$ do not detect homotopy elements. Instead they support nontrivial differentials, namely $d_2(h_j) = h_0 h_{j-1}^2$.
- Browder's theorem [**Bro69**] says there is a framed manifold with non trivial Kervaire invariant in dimension $2^{j+1} - 2$ iff h_j^2 is a permanent cycle, in which case it detects the corresponding homotopy element, θ_j.

Here is a chart showing $\mathrm{Ext}^{s,t}$ for $t - s \leq 20$.

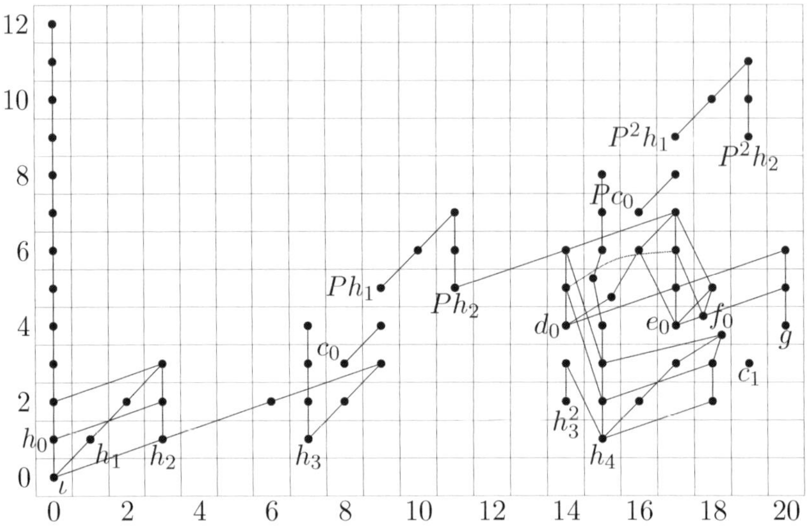

Here lines going up and to the right indicate multiplication by h_0, h_1 and h_2, and lines going up and to the left lines indicate differentials. For more information we refer the reader to [**Rav86**].

4. Spectra and equivariant spectra

In this section we will define ordinary and equivariant spectra, the objects of study in stable homotopy theory and equivariant stable homotopy theory.

4.1. An informal definition of spectra. Informally, a *prespectrum* D is a collection of pointed spaces (spaces equipped with base points that are preserved by all maps in sight) $\{D_n : n \gg 0\}$ with structure maps $\Sigma D_n \to D_{n+1}$. Here ΣX denotes the suspension or double cone on X, wth the cone line through the basepoint $x_0 \in X$ collapsed to be the base point of ΣX. Then we can define

$$\pi_i(D) = \lim_{\to} \pi_{n+i}(D_n) \quad \text{and} \quad H_i(D) = \lim_{\to} H_{n+i}(D_n).$$

A map $\Sigma X \to Y$ is equivalent (adjoint) to a map $X \to \Omega Y$, where ΩY is the loop space of Y, the space of base point preserving map to Y from the circle S^1.

Thus the adjoint of the structure map is a map $D_n \to \Omega D_{n+1}$. A *spectrum* is a prespectrum for which this map is a homeomorphism for each n. We can always get a spectrum $E = \{E_n : n \in \mathbf{Z}\}$ from the prespectrum D by defining

$$E_n = \lim_{\substack{\to \\ k}} \Omega^k D_{n+k}$$

This makes $E_n = \Omega E_{n+1}$ and infinite loop space. The evident map $D \to E$ does not alter homotopy or homology and is thus a weak homotopy equivalence.

A spectrum E can be suspended or desuspended by defining

$$(\Sigma^j E)_n = E_{j+n} \quad \text{for any integer } j,$$

and we have

$$\pi_k(\Sigma^j E) = \pi_{n-j}(E) \quad \text{and} \quad H_k(\Sigma^j E) = H_{n-j}(E).$$

We can make similar statements about looping, defining

$$(\Omega^j E)_n = \Omega^j E_n = E_{n-j},$$

so $\Omega^j E = \Sigma^{-j} E$. Hence a spectrum can be desuspended or delooped any number of times.

Here are two examples:

(i) For a given pointed space X, let $D_n = \Sigma^n X$, the nth iterated suspension of X. Then D is the *suspension prespectrum of* X, denoted by $\Sigma^\infty X$ or (abusively but commonly) simply by X. The homology of $\Sigma^\infty X$ is the reduced homology the space X. Converting to a spectrum as described above does not change its homotopy or homology. When the space X is S^0, we get the sphere spectrum, whose kth homotopy group is π_k^S.

(ii) For an abelian group A, let $K(A, n)$ be the Eilenberg-Mac Lane space (see Sidebar 3) characterized (up to homotopy equivalence) by

$$\pi_k(K(A, n)) = \begin{cases} A & \text{for } k = n \\ 0 & \text{otherwise.} \end{cases}$$

Then $K(A, n)$ is equivalent to $\Omega K(A, n+1)$. We define a spectrum D by $D_n = K(A, n)$ with structure map $\Sigma K(A, n) \to K(A, n+1)$ being the adjoint of this equivalence. In this case E_n is equivalent to D_n. The resulting spectrum is denoted by HA, the Eilenberg-Mac Lane spectrum for A.

4.2. Orthogonal spectra. Let $\mathcal{T}$ denote the category of compactly generated [**ML71**, VII.8] weak Hausdorff [**McC69**, §2] pointed topological spaces. Compactly generated means that a subset is closed iff its intersection with every compact subspace is closed. Weak Hausdorff means that the continuous image of every compact space is closed. These are technical conditions designed to keep us out of trouble, as explained in the references cited.

Following [**May80**, §5] we define an *orthogonal spectrum* to be a functor to this category from an indexing category $\mathscr{I}$. Its objects are are finite dimensional real inner product spaces V. When $V \subset W$, we will denote the orthogonal compliment of V in W by $W - V$.

The category $\mathscr{I}$ is *enriched over* $\mathcal{T}$ **[Kel82]** , which means that the "set" $\mathscr{I}(V,W)$ of morphisms from V to W is actually a pointed topological space, to be defined shortly. $\mathcal{T}$ is enriched over itself; the set of continuous pointed maps $X \to Y$ can be given the compact-open topology with the constant map as base point. A functor of enriched categories $D : \mathscr{I} \to \mathcal{T}$ is required to induce a continuous pointed map from $\mathscr{I}(V,W) \to \mathcal{T}(D_V, D_W)$ for each V and W in $\mathscr{I}$.

To define $\mathscr{I}(V,W)$, let $O(V,W)$ be the (possibly empty) Stiefel manifold of orthogonal embeddings $f : V \to W$. The space

$$\{(f,w) \in O(V,W) \times W : w \in W - f(V)\}$$

is a vector bundle over $O(V,W)$, and $\mathscr{I}(V,W)$ is defined to be its one point compactification or Thom space. As a set it is

$$\mathscr{I}(V,W) = \bigvee_{f \in O(V,W)} S^{W-f(V)} \tag{13}$$

where S^V denotes the one point compactification of V.

In particular, $\mathscr{I}(V,V) = O(V)_+$, the orthogonal group of V with a disjoint base point. Hence the functor D gives a base point preserving action of $O(V)$ on the pointed space D_V. If V and W have the same dimension, then a choice of isomorphism between them leads to identifications of $\mathscr{I}(V,W)$ and $\mathscr{I}(W,V)$ with $O(V)_+$ and to homeomorphisms between D_V and D_W. Hence the topology of the pointed space D_V depends only on the dimension of V.

If $V \subset W$ is a proper subspace, then $\mathscr{I}(V,W) \supset S^{W-V}$ by (13), and we get a structure map $S^{W-V} \wedge D_V \to D_W$ as in 4.1.

4.3. Equivariant orthogonal spectra. Here we define equivariant orthogonal spectra following **[MM02]**.

For a group G, let $\mathcal{T}_G$ denote the category of compactly generated weak Hausdorff pointed topological G-*spaces*, i.e., spaces with an action of G by base point preserving maps. The base point is always fixed by G. When we encounter a G-space X without a fixed point, we adjoin a disjoint base point and denote it by X_+. The groups we will be concerned with in this paper are finite cyclic 2-groups C_{2^n}.

Like $\mathcal{T}$, $\mathcal{T}_G$ is enriched over itself. The space $\mathcal{T}_G(X,Y) = \mathrm{Map}_*(X,Y)$ of *all* pointed maps $X \to Y$ (not just the equivariant ones) is itself a pointed G-space, with the constant map as base point and the action of a group element $\gamma \in G$ on a map $f : X \to Y$ is given by $(\gamma f)(x) = \gamma f(\gamma^{-1}x)$. The fixed point set $\mathcal{T}_G(X,Y)^G$ is the space of equivariant maps.

One could define another category with the same objects as $\mathcal{T}_G$ but enriched instead over $\mathcal{T}$ in which the morphism space is $\mathcal{T}_G(X,Y)^G$. Both enriched categories have underlying ordinary categories in which the spaces of morphisms are replaced by the corresponding sets.

Sidebar 3 Classifying spaces

The join $X * Y$ of topological spaces X and Y is the topological quotient of $X \times I \times Y$ obtained by identifying $(x, 0, y)$ with $(x', 0, y)$ and $(x, 1, y)$ with $(x, 1, y')$ for $x, x' \in X$ and $y, y' \in Y$. (meaning that its homotopy groups vanish below dimension m) and Y is n-connected, then $X * Y$ is $(m + n)$-connected, e.g. $S^m * S^n = S^{m+n+1}$. This construction can be iterated, namely it is possible to define the n-fold join

$$X_0 * X_1 * \cdots * X_n$$

as a quotient of the space

$$\left\{((t_0, \dots, t_n), (x_0, \dots, x_n)) \in I^{n+1} \times \prod_{0 \le i \le n} X_i \colon \sum_{0 \le i \le n} t_i = 1\right\}.$$

When $t_i = 0$, the coordinate x_i is irrelevant. The set of points with $t_n = 0$ is the $(n-1)$-fold join $X_0 * \cdots * X_{n-1}$.

For a topological group G, let E_nG be the n-fold join of G with itself. It is $(n-1)$-connected, even if G itself is not path connected. It is also a G-space, with G acting by left multiplication on the coordinates in G. This action is free; no point is left fixed by any nontrivial element of G. Using the inclusion maps $E_nG \to E_{n+1}G$ (which commute with the G-action) we can let n go to ∞ and obtain a contractible free G-space EG. Its orbit space is denoted by BG and is called the *classifying space of* G. This construction is due to Milnor [**Mil56**]. It has the following properties.

(i) BG is path connected and G is homotopy equivalent to ΩBG, which means that $\pi_i(G) = \pi_{i+1}(BG)$. If G is discrete, then

$$\pi_i(BG) = \begin{cases} G & \text{for } i = 1 \\ 0 & \text{otherwise.} \end{cases}$$

(ii) BG is functorial in G, meaning the a group homomorphism $\phi : G \to H$ induces a map $B\phi : BG \to BH$. If A is a abelian group, the multiplication map $A \times A \to A$ is a homomorphism leading to a map $BA \times BA \to BA$. It is known that this makes BA itself into a topological abelian group, so the classifying space construction can be iterated. If A is discrete, then

$$\pi_i(B^n A) = \begin{cases} A & \text{for } i = n \\ 0 & \text{otherwise,} \end{cases}$$

so $B^n A$ is the Eilenberg-Mac Lane space $K(A, n)$.

(iii) Let G be the n-dimensional orthogonal group $O(n)$ or the n-dimensional unitary group $U(n)$. There is an n-dimensional real [complex] vector bundle $\xi^n_{\mathbf{R}}$ over $BO(n)$ [$\xi^n_{\mathbf{C}}$ over $BU(n)$] with the following universal property: any such bundle over a

Sidebar 3 Continued

paracompact space X is induced by a unique (up to homotopy) from X to the classifying space, and the bundles over X induced by two such maps are isomorphic iff the maps are homotopic. Details can be found in [**MS74**].

We will use an indexing category $\mathscr{I}^G$ whose objects are finite dimensional orthogonal representations V of G. There is an obvious forgetful functor $\mathscr{I}^G \to \mathscr{I}$. We define $\mathscr{I}^G(V, W)$ to be the space $\mathscr{I}(V, W)$ equipped with the evident G-action. Its fixed point space is

$$\mathscr{I}^G(V, W)^G = \mathscr{I}(V^G, W^G) \wedge O(V^\perp, W^\perp)^G_+,$$

where $V^\perp$ denotes the orthogonal compliment of the invariant subspace V^G of V.

We define an *equivariant orthogonal G-spectrum* to be a functor

$$E : \mathscr{I}^G \to \mathcal{T}_G$$

enriched over G-spaces and equivariant maps. This means that for each representation V we get a pointed G-space E_V, and for each V and W we get a continuous pointed equivariant map

$$\mathscr{I}^G(V, W) \to \mathcal{T}_G(E_V, E_W).$$

As before we have structure maps

$$S^V \wedge E_W \to E_{V+W}$$

which are equivariant. Here S^V is the one point compactification of the representation V, which has its own action of G determined by the representation.

We denote the category of equivariant orthogonal G-spectra by $\mathcal{S}^G$. It is an enriched functor category [**ML71**, II.4] in which objects are continuous functors $\mathscr{I}^G \to \mathcal{T}_G$ as described above, and morphisms are natural transformations. Thus a morphism of equivariant orthogonal G-spectra $g : D \to E$ consists of continuous equivariant maps $g_V : D_V \to E_V$ for each $V \in \mathscr{I}^G$ such that for each $V, W \in \mathscr{I}^G$ the following diagram commutes

$$\begin{array}{ccc} \mathscr{I}^G(V, W) & \xrightarrow{\quad D \quad} & \mathcal{T}_G(D_V, D_W) \\ \downarrow{\scriptstyle E} & & \downarrow{\scriptstyle \mathcal{T}_G(D_V, g_W)} \\ \mathcal{T}_G(E_V, E_W) & \xrightarrow{\mathcal{T}_G(g_V, E_W)} & \mathcal{T}_G(D_V, E_W). \end{array}$$

Here the vertical and horizontal arrows pointing toward $\mathcal{T}_G(D_V, E_W)$ are respectively composition with g_W and precomposition with g_V.

For each virtual representation $U-W$ there is a spectrum S^{U-W} defined by

$$(S^{U-W})_V = \mathcal{T}(S^W, S^{U+V}).$$

The group of homotopy classes of equivariant maps $\pi^G_{V-W}(X) = [S^{V-W}, X]$ depends only on the isomorphism class of $V - W$, which is an element in the orthogonal representation ring $RO(G)$. *Thus we get homotopy groups graded over $RO(G)$ rather than the integers.* These groups are collectively denoted by $\pi^G_\star(X)$, with the five pointed star $\star$ in place of the usual six pointed asterisk $*$.

For any subgroup $H \subset G$, there is a restriction functor $i^G_H : \mathcal{S}^G \to \mathcal{S}^H$, so we also have homotopy groups $\pi^H_\star(X)$ graded over $RO(H)$. In the case of the trivial subgroup we denote this by $\pi^u_*(X)$, the homotopy of the underlying (nonequivariant) spectrum of X. For each subgroup H, when the subscript is an ordinary integer (meaning a trivial representation of H), we are looking at H-equivariant maps from spheres fixed by H, so we have

$$\pi^H_*(X) = \pi_*(X^H)$$

where X^H is the fixed point spectrum of H.

References

[Ada58] J. F. Adams. On the structure and applications of the Steenrod algebra. *Comment. Math. Helv.*, 32:180–214, 1958.

[Ada60] J. F. Adams. On the non-existence of elements of Hopf invariant one. *Ann. of Math. (2)*, 72:20–104, 1960.

[Ada62] J. F. Adams. Vector fields on spheres. *Ann. of Math. (2)*, 75:603–632, 1962.

[Ada66] J. F. Adams. On the groups $J(X)$. IV. *Topology*, 5:21–71, 1966.

[Arf41] Cahit Arf. Untersuchungen über quadratische Formen in Körpern der Charakteristik 2. I. *J. Reine Angew. Math.*, 183:148–167, 1941.

[BJM84] M. G. Barratt, J. D. S. Jones, and M. E. Mahowald. Relations amongst Toda brackets and the Kervaire invariant in dimension 62. *J. London Math. Soc. (2)*, 30(3):533–550, 1984.

[BMT70] M. G. Barratt, M. E. Mahowald, and M. C. Tangora. Some differentials in the Adams spectral sequence. II. *Topology*, 9:309–316, 1970.

[Bot59] R. Bott. The stable homotopy of the classical groups. *Annals of Mathematics*, 70:313–337, 1959.

[BP66] Edgar H. Brown, Jr. and Franklin P. Peterson. The Kervaire invariant of $(8k+2)$-manifolds. *Amer. J. Math.*, 88:815–826, 1966.

[Bro69] William Browder. The Kervaire invariant of framed manifolds and its generalization. *Ann. of Math. (2)*, 90:157–186, 1969.

[Fre38] H. Freudenthal. Über die Klassen der Sphärenabbildungen I. Grosse Dimnesionionen . *Compositio Math.*, 5:299–314, 1938.

[GM95] J. P. C. Greenlees and J. P. May. Equivariant stable homotopy theory. In *Handbook of algebraic topology*, pages 277–323. North-Holland, Amsterdam, 1995.

[Gra75] Brayton Gray. *Homotopy theory.* Academic Press [Harcourt Brace Jovanovich Publishers], New York, 1975. An introduction to algebraic topology, Pure and Applied Mathematics, Vol. 64.

[Hat02] Allen Hatcher. *Algebraic topology.* Cambridge University Press, Cambridge, 2002.

[Haz78] Michiel Hazewinkel. *Formal groups and applications*, volume 78 of *Pure and Applied Mathematics*. Academic Press Inc. [Harcourt Brace Jovanovich Publishers], New York, 1978.

[Hop30] H. Hopf. Über die Abbildungen der dreidimensionalen Sphäre auf die Kugelfläche. *Mathematische Annalen*, 104:639–665, 1930.

[Hop35] H. Hopf. Über die Abbildungen von Sphären auf Sphären niedrigerer Dimension. *Fundamenta Mathematica*, 25:427–440, 1935.

[Hur35] W. Hurewicz. Beiträge zur Topologie Deformationen I–II. *Nederl. Akad. Wetensch. Proc., Series A*, 38:112–119, 521–528, 1935.

[Hur36] W. Hurewicz. Beiträge zur Topologie Deformationen III–IV. *Nederl. Akad. Wetensch. Proc., Series A*, 39:117–126, 215–224, 1936.

[Jam55] I. M. James. Reduced product spaces. *Ann. of Math. (2)*, 62:170–197, 1955.

[Jam56a] I. M. James. On the suspension triad. *Ann. of Math. (2)*, 63:191–247, 1956.

[Jam56b] I. M. James. The suspension triad of a sphere. *Ann. of Math. (2)*, 63:407–429, 1956.

[Jam57] I. M. James. On the suspension sequence. *Ann. of Math. (2)*, 65:74–107, 1957.

[Jon78] John D. S. Jones. The Kervaire invariant of extended power manifolds. *Topology*, 17(3):249–266, 1978.

[Kel82] Gregory Maxwell Kelly. *Basic concepts of enriched category theory*, volume 64 of *London Mathematical Society Lecture Note Series*. Cambridge University Press, Cambridge, 1982.

[Ker60] Michel A. Kervaire. A manifold which does not admit any differentiable structure. *Comment. Math. Helv.*, 34:257–270, 1960.

[Mah67] Mark Mahowald. *The metastable homotopy of S^n*. Memoirs of the American Mathematical Society, No. 72. American Mathematical Society, Providence, R.I., 1967.

[Mah82] Mark Mahowald. The image of J in the EHP sequence. *Ann. of Math. (2)*, 116(1):65–112, 1982.

[May80] J. P. May. Pairings of categories and spectra. *J. Pure Appl. Algebra*, 19:299–346, 1980.

[May99] J. P. May. *A concise course in algebraic topology*. Chicago Lectures in Mathematics. University of Chicago Press, Chicago, IL, 1999.

[McC69] M. C. McCord. Classifying spaces and infinite symmetric products. *Trans. Amer. Math. Soc.*, 146:273–298, 1969.

[Mil56] John Milnor. Construction of universal bundles. II. *Ann. of Math. (2)*, 63:430–436, 1956.

[Mil58] John Milnor. The Steenrod algebra and its dual. *Ann. of Math. (2)*, 67:150–171, 1958.

[Mil59] John Milnor. On spaces having the homotopy type of a CW-complex. *Trans. Amer. Math. Soc.*, 90:272–280, 1959.

[Mil60] J. Milnor. On the cobordism ring Ω^* and a complex analogue. I. *Amer. J. Math.*, 82:505–521, 1960.

[ML71] Saunders Mac Lane. *Categories for the working mathematician*. Springer-Verlag, New York, 1971. Graduate Texts in Mathematics, Vol. 5.

[MM02] M. A. Mandell and J. P. May. Equivariant orthogonal spectra and S-modules. *Mem. Amer. Math. Soc.*, 159(755):x+108, 2002.

[MS74] John W. Milnor and James D. Stasheff. *Characteristic classes*. Princeton University Press, Princeton, N. J., 1974. Annals of Mathematics Studies, No. 76.

[MT67] Mark Mahowald and Martin Tangora. Some differentials in the Adams spectral sequence. *Topology*, 6:349–369, 1967.

[Nov60] S. P. Novikov. Some problems in the topology of manifolds connected with the theory of Thom spaces. *Soviet Math. Dokl.*, 1:717–720, 1960.

[Nov62] S. P. Novikov. Homotopy properties of Thom complexes. *Mat. Sb. (N.S.)*, 57 (99):407–442, 1962.

[Nov67] S. P. Novikov. Methods of algebraic topology from the point of view of cobordism theory. *Izv. Akad. Nauk SSSR Ser. Mat.*, 31:855–951, 1967.

[Pon38] L. S. Pontryagin. A classification of continuous transformations of a complex into a sphere, 1 and 2. *Doklady Akad. Nauk SSSR (N.S.)*, 19:147–149 and 361–363, 1938.

[Pon50] L. S. Pontryagin. Homotopy classification of the mappings of an $(n+2)$-dimensional sphere on an n-dimensional one. *Doklady Akad. Nauk SSSR (N.S.)*, 70:957–959, 1950.

[Pon55] L. S. Pontryagin. *Gladkie mnogoobraziya i ih primeneniya v teorii gomotopiĭ[Smooth manifolds and their applications in homotopy theory]*. Trudy Mat. Inst. im. Steklov. no. 45. Izdat. Akad. Nauk SSSR, Moscow, 1955.

[Qui69] Daniel Quillen. On the formal group laws of unoriented and complex cobordism theory. *Bull. Amer. Math. Soc.*, 75:1293–1298, 1969.

[Rav86] Douglas C. Ravenel. *Complex cobordism and stable homotopy groups of spheres*, volume 121 of *Pure and Applied Mathematics*. Academic Press Inc., Orlando, FL, 1986. Errata and second edition available online at author's home page.

[Rav04] Douglas C. Ravenel. *Complex Cobordism and Stable Homotopy Groups of Spheres, Second Edition*. American Mathematical Society, Providence, 2004. Available online at `www.math.rochester.edu/u/faculty/doug/mu.html`.

[Ser51] Jean-Pierre Serre. Homologie singulière des espaces fibrés. Applications. *Ann. of Math. (2)*, 54:425–505, 1951.

[Sna09] Victor P. Snaith. *Stable homotopy around the Arf-Kervaire invariant*, volume 273 of *Progress in Mathematics*. Birkhäuser Verlag, Basel, 2009.

[Ste62] N. E. Steenrod. *Cohomology operations*. Lectures by N. E. Steenrod written and revised by D. B. A. Epstein. Annals of Mathematics Studies, No. 50. Princeton University Press, Princeton, N.J., 1962.

[Ste99] Norman Steenrod. *The topology of fibre bundles*. Princeton Landmarks in Mathematics. Princeton University Press, Princeton, NJ, 1999. Reprint of the 1957 edition, Princeton Paperbacks.

[Sto68] Robert E. Stong. *Notes on Cobordism Theory*. Princeton University Press, Princeton, 1968.

[Tho54] René Thom. Quelques propriétés globales des variétés différentiables. *Comment. Math. Helv.*, 28:17–86, 1954.

[Whi42] G. W. Whitehead. On the homotopy groups of spheres and rotation groups. *Annals of Mathematics*, 43:634–640, 1942.

[Whi49] J. H. C. Whitehead. Combinatorial homotopy. I. *Bull. Amer. Math. Soc.*, 55:213–245, 1949.

[Whi50] George W. Whitehead. The $(n+2)^{\mathrm{nd}}$ homotopy group of the n-sphere. *Ann. of Math. (2)*, 52:245–247, 1950.

Department of Mathematics, University of Virginia, Charlottesville, VA 22904

E-mail address: `michael.a.hill@math.uva.edu`

Department of Mathematics, Harvard University, Cambridge, MA 02138

E-mail address: `mjh@math.harvard.edu`

Department of Mathematics, University of Rochester, Rochester, NY 14627

E-mail address: `doug+pi@math.rochester.edu`

Current Developments in Mathematics, 2009

Phase Transitions, Minimal Surfaces and A Conjecture of De Giorgi[1]

O. Savin

1. Introduction

A central problem in the area of PDEs is the study of global solutions, that is solutions defined in the whole space. This problem arises naturally for example when one studies the possible types of behaviors a solution might have at a given point. Focusing near such a point by dilating the picture more and more, we end up with a PDE in the whole space. In other cases we might have a differential equation with oscillating terms on a small scale depending on a parameter ε. In order to understand the behavior at unit scale as the parameter $\varepsilon \to 0$ we often need to dilate the picture and solve some PDE in the whole space. As a consequence, the study of global solutions is crucial when dealing with singularities, local behavior and homogenization limits of PDEs.

In this paper we present symmetry results for global solutions to certain semilinear (or fully nonlinear) elliptic equations of the type

$$\triangle u = f(u)$$

which arise in the theory of phase transitions.

Let us briefly introduce the typical physical model for a phase transition. Imagine in a domain Ω we have a two-phase fluid whose density at a point x we denote by $\rho(x)$. Assume its energy density is given by a double-well potential $W(\rho(x))$ say with minima at ρ_1, ρ_2 i.e.

$$W(\rho_1) = W(\rho_2) = 0, \quad W(s) > 0 \quad \text{if } s \neq \rho_1, \rho_2.$$

The densities ρ_1 and ρ_2 correspond to the stable fluid phases. Then the energy of the fluid is given by the integral

$$\int_\Omega W(\rho(x))\, dx.$$

[1]The author was supported by N.S.F. Grant DMS-07-01037 and a Sloan Fellowship.

This is not a satisfactory physical model since any density function $\rho(x)$ that takes only the values ρ_1 and ρ_2 minimizes the density energy. In particular the stable phases ρ_1 and ρ_2 could coexist along any complicated interface! The reason for this is because we ignored the kinetic energy term that takes into consideration interactions at small scales (such as friction) which penalizes the formation of unnecessary interfaces. The energy associated with the fluid in this case would be of the form

$$J_\varepsilon(\rho) = \int_\Omega \frac{\varepsilon^2}{2}|\nabla \rho|^2 + W(\rho)\, dx$$

with ε a small parameter. This is the typical energy modeling the phase separation phenomena within the van der Walls-Cahn-Hilliard theory (see [**12**]).

The presence of the gradient square term in the energy prevents instantaneous jumps from a region of density ρ_1 to a region of density ρ_2, and in general this stable states are only asymptotically attained by ρ. The transition region between the two phases occurs in a thin region of width comparable to ε. If we try to understand this transition at length-scale ε we need to dilate the picture by a factor of $1/\varepsilon$. Then the rescaled density $u(x) := \rho(\varepsilon x)$ minimizes the energy

$$\int \frac{1}{2}|\nabla u|^2 + W(u)\, dx$$

and hence solves the Euler-Lagrange equation

$$\triangle u = W'(u),$$

in the large domain Ω/ε. On the other hand, we shall see in Section 2 that if we let $\varepsilon \to 0$ in the original domain Ω, the transition region converges to a minimal surface inside Ω, i.e. a surface of least area. This fact was already known to the school of De Giorgi in Pisa in the late 1970's, and De Giorgi conjectured that global solutions to the Euler-Lagrange equation above should have similar properties to something apparently unrelated, global minimal surfaces.

The goal of this paper is to present several results in both settings, phase transitions and minimal surfaces, together with their common underlying ideas. Special interest is given to flatness theorems and their application to

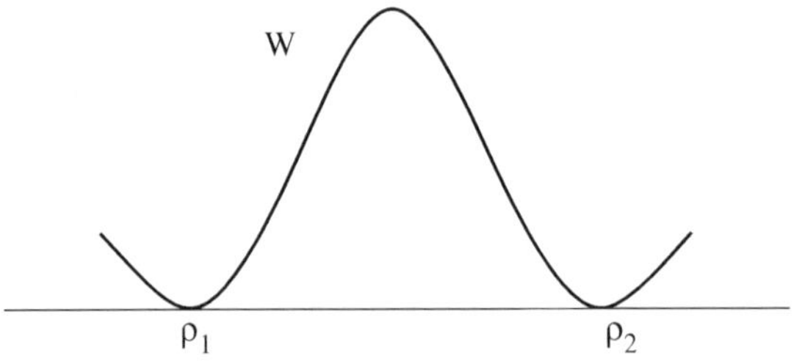

FIGURE 1. Double-well potential

1D symmetry of global phase transitions and in particular to De Giorgi's conjecture. See Section 3 for the precise formulation of this conjecture.

We conclude our introduction with a brief description of the theory of minimal surfaces, with special emphasis on the ideas of De Giorgi.

Finding the surface of least area (minimal surface) among all surfaces bounded by a given curve is known as the Plateau problem. It is one of the oldest problems in mathematics and it has received a great deal of attention from the mathematical community. Plateau problem was solved independently by Douglas [**27**] and Rado [**53, 54**] in the 1930's. Later, the theories of De Giorgi on perimeters [**21**–**24**], the one of Federer and Fleming [**33, 35**] on rectifiable currents and the one of Almgreen [**2, 3**] on varifolds were developed to deal with the higher dimensional case.

Let us explain the approach of De Giorgi. The key idea is to look at hypersurfaces in $\mathbb{R}^n$ as boundaries of sets. Precisely, for any measurable set E, we can define the perimeter of E in a domain $\Omega \subset \mathbb{R}^n$ (or the area of ∂E in Ω) as the total variation of the characteristic function of E, χ_E, in Ω

$$P_\Omega(E) = \int_\Omega |\nabla \chi_E| = \sup_{|g|\leq 1} \left| \int_E div\, g\, dx \right|,$$

where the supremum is taken over all vector fields $g \in C_0^1(\Omega)$ (with compact support in Ω) with $|g| \leq 1$. From Green's theorem we see that this notion of area of ∂E coincides with the usual one whenever ∂E is a C^1 hypersurface. The advantage of defining the perimeter for a larger class of sets is the compactness in L^1 of sets with finite perimeter. Then it is not difficult to show existence to the Plateau problem in this context of minimal boundaries. It is much more difficult to prove that the sets so obtained are actually smooth hypersurfaces except possibly for a closed singular set of small Hausdorff dimension.

One understands the behavior of such a minimal surface in a neighborhood of a point x_0 by using the blow-up technique explained at the beginning. Dilating the picture more and more we end up with a limiting minimal surface defined in the whole space. What can be said about global minimal surfaces? Simons [**59**] proved that the only global minimal surfaces are the hyperplanes in dimension $n \leq 7$. Moreover, Bombieri, De Giorgi and Giusti [**11**] showed that the symmetric cone in $\mathbb{R}^8$

$$\{x \in \mathbb{R}^8 : x_1^2 + x_2^2 + x_3^2 + x_4^2 = x_5^2 + x_6^2 + x_7^2 + x_8^2\}$$

known as Simons cone is minimal. This is the first example of a minimal surface that has a singularity (at the origin). Going back to the original picture, Simons result says that at least in low dimensions the minimal surface is approximated well by a plane (it is flat) in a small neighborhood of x_0. On the other hand, a theorem of De Giorgi that holds in any dimension says that flat minimal surfaces in $B_r(x_0)$ are analytic in $B_{r/2}(x_0)$. This "flatness theorem" gives smoothness near the points that have a planar blow-up limit. Thus, minimal surfaces are always smooth in dimension $n \leq 7$.

In higher dimensions, it was shown by Federer [**34**] that they are smooth everywhere except on a small singular set of Hausdorff dimension $n-8$.

Minimal surfaces that are graphs in some direction are called minimal graphs. It turns out that minimal graphs are smooth in any dimension. The question wether the only global minimal graphs are planar is known as the Bernstein problem, due to Bernstein [**10**] (1915) which answered affirmatively the question in $\mathbb{R}^3$. Bombieri, De Giorgi and Giusti [**11**] showed that the graph assumption gains one extra dimension with respect to the minimal surface results mentioned earlier: minimal graphs are planar in dimension $n \le 8$, and there are non-planar minimal graphs in $\mathbb{R}^9$.

The paper is organized as follows. In Section 2 we give the precise relation between minimal surfaces and level sets of minimizers of J_ε. In Section 3 we state the conjecture of De Giorgi together with other related problems. In Section 4 we prove density estimates in both settings. In Section 5 and 6 we give a new proof of the flatness theorem of De Giorgi for minimal surfaces which is based on Harnack inequality. In Sections 7 to 9 we extend these ideas to phase transitions and obtain the corresponding flatness theorem. In Section 10 and 11 we discuss $1D$ symmetry for global phase transitions. Finally in Section 12 we present similar results that are available for fully nonlinear equations.

2. Phase transitions and minimal surfaces

We consider the Ginzburg-Landau energy functional

$$J(u,\Omega) = \int_\Omega \frac{1}{2}|\nabla u|^2 + W(u)\, dx, \quad |u| \le 1$$

with W a double-well potential with minima at 1 and -1 satisfying

$$W \in C^2([-1,1]), \quad W(-1) = W(1) = 0, \quad W > 0 \text{ on } (-1,1),$$
$$W'(-1) = W'(1) = 0, \quad W''(-1) > 0, \quad W''(1) > 0.$$

The classical double-well potential W to have in mind is

$$W(s) = \frac{1}{4}(1-s^2)^2.$$

We mention that all the results of this paper apply also to more general potentials W that do not necessarily satisfy the conditions above. Such an example is the discontinuous potential $W(s) = \chi_{(-1,1)}(s)$, where $\chi_{(-1,1)}$ represents the characteristic function of the interval $(-1,1)$. In this case the physical motivation comes from a problem associated with jet fluid flows (see [**4**]).

In this section we explain the close relation between minimal surfaces and level sets of minimizers of J.

Consider a minimizer u of J which is bounded by -1 and 1. We say that u is a minimizer in the (possible unbounded) set Ω if its energy increases locally under any compact perturbation, i.e. for any open set A relatively compact in Ω,

$$J(u, A) \leq J(u + \varphi, A) \quad \forall \varphi \in H_0^1(A).$$

A minimizer in whole $\mathbb{R}^n$ is called a global minimizer.

The behavior of u in large domains is given by the behavior of the rescaled functions u_ε in B_1,

$$u_\varepsilon(x) := u(\frac{x}{\varepsilon}).$$

If u minimizes J in the ball $B_{\frac{1}{\varepsilon}}$ then u_ε minimizes the rescaled energy J_ε in B_1

$$J_\varepsilon(v, B_1) := \int_{B_1} \frac{\varepsilon}{2}|\nabla v|^2 + \frac{1}{\varepsilon}W(v)\,dx.$$

We continue with a heuristic discussion about minimizers of J_ε. For a given function v, the main contribution in $J_\varepsilon(v)$ comes from the potential energy which is minimized when v equals either 1 or -1. Instant jumps from a region where $v = 1$ to a region where $v = -1$ are not allowed since the kinetic energy $\int \frac{\varepsilon}{2}|\nabla v|^2$ would become infinite. However, it is useful to notice that if instead of the Dirichlet integral we would have the BV norm $\int \varepsilon|\nabla v|$ then such jumps are allowed and in this case the energy is minimized when the jumps occur along a minimal surface.

Clearly

$$\int_{B_1} \frac{\varepsilon}{2}|\nabla v|^2 + \frac{1}{\varepsilon}W(v)\,dx \geq \int_{B_1} \sqrt{2W(v)}|\nabla v|\,dx,$$

and using the co-area formula

$$|\nabla v|\,dx = d_{\mathcal{H}^{n-1}(\{v=s\})}ds$$

the inequality becomes

$$J_\varepsilon(v) \geq \int_{-1}^{1} \sqrt{2W(s)}\mathcal{H}^{n-1}(\{v = s\})\,ds.$$

The energy J_ε is then minimized by the function v if every level set is a minimal surface and we have equality in the inequality above i.e.

$$|\nabla v| = \frac{1}{\varepsilon}\sqrt{2W(v)}.$$

This last equality gives

$$v(x) = g_0\left(\frac{d_\Gamma(x)}{\varepsilon}\right),$$

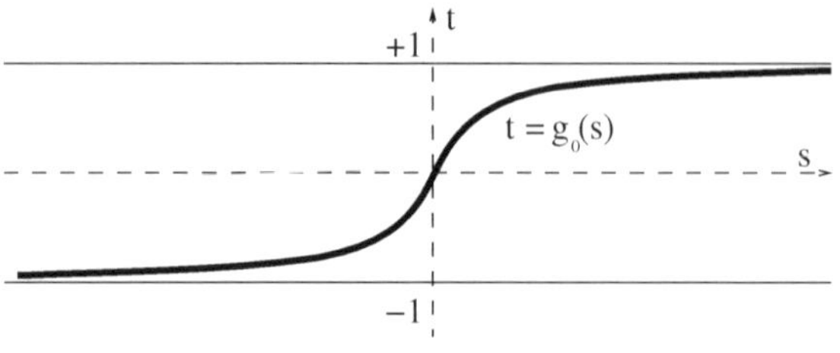

FIGURE 2. One-dimensional solution

where $d_\Gamma(x)$ represents the sign distance to the 0 level set $\Gamma := \{v = 0\}$ and g_0 is the solution to the ODE

$$g_0' = \sqrt{2W(g_0)}, \quad g_0(0) = 0.$$

The function g_0 is in fact the unique minimizer of J in 1D which is increasing.

In general the level sets of a function v as above cannot be all minimal surfaces. However, if for example the 0 level set Γ is minimal then the s-level sets are essentially minimal as long as s is not too close to ± 1 and ε is small. On the other hand when s is close to ± 1 the weight $\sqrt{2W(s)}ds$ becomes negligible. Thus such a function is "almost" a minimizer for J_ε. This suggests that the level sets of minimizers of J_ε converge to a minimal surface as $\varepsilon \to 0$.

The asymptotic behavior of u_ε was first studied by Modica and Mortola [**51**] and by Modica [**45**] within the framework of Γ-convergence. Later Modica [**47**], Sternberg [**60**] and many authors [**6, 36, 43, 49, 52, 61**] generalized these results for minimizers with volume constraint.

Here we present the result of Modica in [**45**].

THEOREM 2.1 (Modica). *Let u_ε be minimizers for the energies $J_\varepsilon(\cdot, B_1)$. There exists a sequence u_{ε_k} such that*

$$u_{\varepsilon_k} \to \chi_E - \chi_{E^c} \quad \text{in } L^1_{loc}(B_1)$$

where E is a set with minimal perimeter in B_1. Moreover, if A is an open set, relatively compact in B_1, such that

$$\int_{\partial A} |D\chi_E| = 0,$$

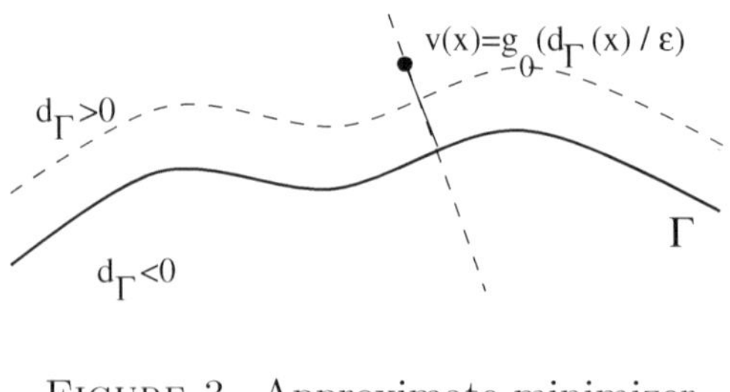

FIGURE 3. Approximate minimizer

then

$$\lim_{m\to\infty} J_{\varepsilon_k}(u_{\varepsilon_k}, A) = P_A(E)\int_{-1}^{1}\sqrt{2W(s)}ds. \tag{2.1}$$

Let us recall the notion of perimeter $P_\Omega(E)$ and minimal perimeter that appear in the theorem above. The perimeter of E in $\Omega \subset \mathbb{R}^n$ is defined as

$$P_\Omega(E) = \sup\left|\int_E div\, g\, dx\right|,$$

where the supremum is taken over all vector fields $g \in C_0^1(\Omega)$ with $\|g\| \leq 1$. When Ω is the whole $\mathbb{R}^n$ we use the shorter notation

$$P(E) = P_{\mathbb{R}^n}(E).$$

We say that E is a set with minimal perimeter in Ω or shortly ∂E is a minimal surface in Ω if, for every open set $A \subset \Omega$, relatively compact in Ω

$$P_A(E) \leq P_A(F),$$

whenever E and F coincide outside a compact set included in A.

From the heuristic discussion before the theorem of Modica we see that the converse is also true at least in the case when the minimal surface ∂E is unique i.e. given a minimal surface ∂E we can find a sequence of minimizers u_ε converging to $\chi_E - \chi_{CE}$. Therefore minimal surfaces and minimizers of J_ε should have similar properties at least for small ε.

In Section 4 we obtain density estimates for level sets of minimizers. As a consequence, the convergence of u_{ε_k} in Theorem 2.1 is in fact stronger than L^1_{loc} i.e. the level sets of u_{ε_k} converge uniformly on compact sets to ∂E.

If the functions u_{ε_k} are all rescalings of a global minimizer u of J, then the blow-down sets

$$\{u_{\varepsilon_k} = 0\} = \varepsilon_k\{u = 0\}$$

converge uniformly to a minimal surface. It is a difficult problem to understand how well the original level set $\{u = 0\}$ can be approximated by a minimal surface. For example in the case when $\partial E = \{x_n = 0\}$ it is not clear wether or not $\{u = 0\}$ stays at a bounded distance from ∂E. Conversely,

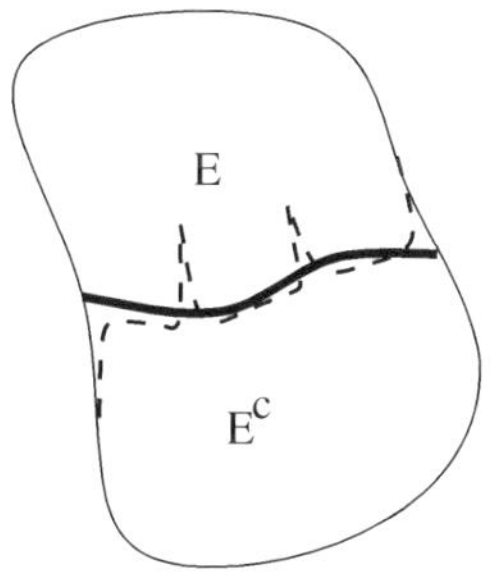

FIGURE 4. Convergence of u_ε.

given a nontrivial global minimal surface ∂E it is not evident if there exists at least one minimizer for which $\{u = 0\}$ stays at a bounded distance from ∂E. Such an example was given by Del Pino, Kowalczyk and Wei [**26**] in dimension 9 which is the first dimension for which a nontrivial global minimal graph exists.

3. De Giorgi's conjecture

In view of the connection between minimal surfaces and minimizers of J, De Giorgi [**20**] made in 1978 the following conjecture about bounded solutions of a global semilinear equation:

CONJECTURE (De Giorgi). *Let $u \in C^2(\mathbb{R}^n)$ be a solution of*

$$\triangle u = u^3 - u,$$

such that

$$|u| < 1, \quad u_{x_n} > 0$$

in whole $\mathbb{R}^n$. Is it true that all the level sets of u are hyperplanes, at least if $n \leq 8$?

The conjecture is sometimes referred as "the ε version of the Bernstein problem for minimal graphs". This relation with the Bernstein problem is the reason why $n \leq 8$ appears in the conjecture.

De Giorgi's conjecture is often considered with the additional natural hypothesis which is coherent with the phase transition setting

$$\lim_{x_n \to \pm\infty} u(x', x_n) = \pm 1. \tag{3.1}$$

The monotonicity $u_{x_n} > 0$ implies that the level sets of u are graphs in the e_n direction whereas the condition (3.1) says that these graphs are defined over whole $\mathbb{R}^{n-1}$. Under the much stronger assumption that the limits in (3.1) are uniform in x', the conjecture is known as the *Gibbons conjecture*. In this case the level sets of u are bounded graphs in the x_n direction.

Gibbons conjecture was first proved for $n \leq 3$ by Ghoussoub and Gui in [**37**] and then for all dimensions n independently by Barlow, Bass and Gui [**7**], Berestycki, Hamel and Monneau [**9**] and Farina [**29**].

The first positive partial result on the De Giorgi conjecture was established in 1980 by Modica and Mortola [**50**]. They proved the conjecture in dimension $n = 2$ under the additional hypothesis that the level sets $\{u = s\}$ are equi-Lipschitz in the x_2 direction. Their proof used a Liouville-type theorem for elliptic equations in divergence form, due to Serrin, for the bounded ratio

$$\sigma := \frac{u_{x_1}}{u_{x_2}}.$$

In 1997 Ghoussoub and Gui [**37**] proved De Giorgi conjecture for $n = 2$. They used a different Liouville-type theorem for σ developed by Berestycki, Caffarelli and Nirenberg [**8**] for the study of symmetry properties of positive solutions of semilinear elliptic equations in half spaces. This theorem does not require for σ to be bounded, but rather a compatibility condition between the growth of σ and the degeneracy of the coefficients of the equation. Using similar techniques, Ambrosio and Cabre [**5**] extended these results to dimension $n = 3$. Also, Ghoussoub and Gui showed in [**38**] that the conjecture is true for $n = 4$ or $n = 5$ for a special class of solutions that satisfy an anti-symmetry condition. We shall discuss the methods described above in Section 11.

In 2003 the conjecture was proved by the author in [**55**] for $n \leq 8$ under the additional hypothesis (3.1). One of the goals of this expository paper is to explain the main ideas of [**55**] (see Sections 7-10). The methods developed do not depend so much on the particular dimension but rather they allow one to obtain rigidity results for global phase transitions as a consequence of the results for minimal surfaces.

Recently, Del Pino Kowalczyk and Wei [**26**] gave a counterexample to the De Giorgi's conjecture in dimension 9. This counterexample satisfies also the limiting assumption (3.1). The construction is based on careful perturbations of the level sets around the nonplanar minimal graph in $\mathbb{R}^9$ constructed by De Giorgi, Bombieri and Giusti.

De Giorgi's conjecture refers to one analogy between phase transitions and minimal surfaces, i.e. the Bernstein problem. There are many other similar questions that can be posed. For example, is it true that minimizers of J are one-dimensional at least for $n \leq 7$? If the level sets are Lipschitz graphs, is it true that the solution must be one-dimensional in any dimension? What if the level sets satisfy only a one-sided bound? What happens if we consider a fully nonlinear equation instead of the semilinear one? We answer these questions in Section 10-12.

Referring to the questions above, when the level sets are assumed to be Lipschitz graphs, it was first shown by Barlow, Bass and Gui [**7**] that u is one-dimensional with probabilistic methods. Different proofs were given by the author in [**55**] and by Caffarelli and Cordoba in [**18**]. Also, we mention that Cabre and Terra in [**13**] are investigating the existence of nontrivial symmetric global minimizers in dimension 8, which are analogous to Simons cone.

There are further similarities between the theory of phase transitions and the theory of minimal surfaces besides the rigidity of global solutions. For example level sets of minimizers of J satisfy uniform density estimates (see next section). Also, there is a monotonicity formula for minimizers which is due to Modica (see [**48**]) and reads

$$\Phi(r) = \frac{J_{B_r}(u)}{r^{n-1}}$$

is increasing in r. However, this monotonicity formula is not as powerful as the one in the minimal surface case since Φ is not scale invariant and it is not constant for the one-dimensional solutions.

Caffarelli and Cordoba developed in [**17**] a theory for phase transitions similar to the one of De Giorgi for Lipschitz minimal graphs. Also, as we will see in Section 7, phase transitions satisfy an improvement of flatness theorem which is key in our analysis and corresponds to De Giorgi flatness theorem for minimal surfaces.

4. Density estimates

In this section we prove density estimates for minimal surfaces and phase transitions. These are crucial in both theories since they allow us to pass from convergence in measure (L^1 convergence) to uniform convergence.

4.1. Minimal surfaces. Let E be a measurable set in $\mathbb{R}^n$.

We say that $0 \in \partial E$ (in measure sense) if for any $\varepsilon > 0$ we have

$$|B_\varepsilon \cap E| > 0, \quad |B_\varepsilon \cap \mathcal{C}E| > 0,$$

where $\mathcal{C}E$ denotes the complement of E and $|E|$ the Lebesgue measure of E. It is not difficult to see that ∂E in measure sense is a closed set.

THEOREM 4.1 (Density estimates). *Assume that E has minimal perimeter in B_1 and $0 \in \partial E$. There exists a constant $c > 0$ depending only on the dimension n such that for all $r \in (0, 1)$*

$$|E \cap B_r| > cr^n, \quad |\mathcal{C}E \cap B_r| > cr^n.$$

PROOF. Minimality implies

$$P_{B_1}(E) \leq P_{B_1}(E \setminus B_r)$$

or

$$P_{B_r}(E) \leq \mathcal{H}^{n-1}(E \cap \partial B_r)$$

hence

$$P(E \cap B_r) \leq 2\mathcal{H}^{n-1}(E \cap \partial B_r).$$

We denote $V(r) = |B_r \cap E|$ and we use the isoperimetric inequality

$$cV(r)^{\frac{n-1}{n}} \leq P(E \cap B_r)$$

to obtain

$$cV(r)^{\frac{n-1}{n}} \leq V'(r)$$

or

$$c \leq (V^{\frac{1}{n}}(r))'.$$

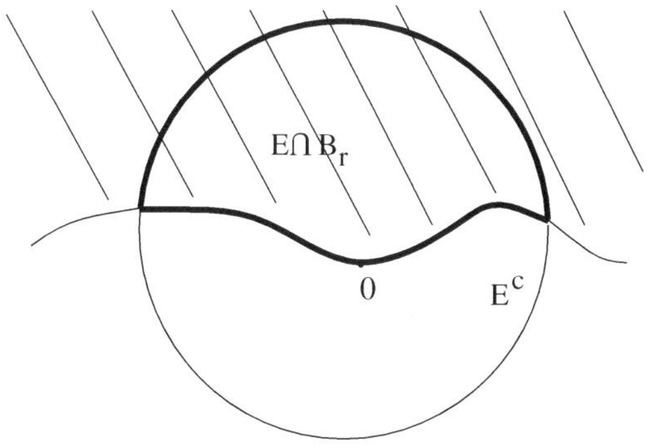

FIGURE 5. Density estimate

Here we used that $V(r) > 0$ for all $r > 0$ since $0 \in \partial E$. We integrate the last inequality and obtain the first density estimate. The second estimate follows by considering $\mathcal{C}E$. □

Next we recall the standard compactness theorem for sets with minimal perimeter. A proof can be found in the book of Giusti [**40**].

THEOREM 4.2 (Compactness). *If E_n is a sequence of minimal sets in Ω then there exists a subsequence E_{n_k} that converges to a minimal set E i.e.*

$$\chi_{E_{n_k}} \to \chi_E \quad in\ L^1_{loc}(\Omega).$$

The density estimates imply that the minimal surfaces ∂E_k converge in Hausdorff sense to ∂E on any compact set of Ω. Indeed, if for some large k there exists a point $x_0 \in \partial E_k$ at distance δ from ∂E, then from the density estimates we find

$$\int_{B_\delta(x_0)} |\chi_{E_k} - \chi_E| \geq c|B_\delta|,$$

and we contradict the convergence of E_k to E in L^1_{loc}.

As we explained in the introduction, an important example occurs when $\Omega = \mathbb{R}^n$ and the limiting minimal surface ∂E is planar, say $\partial E = \{x_n = 0\}$. Then, the uniform convergence implies that for any $\varepsilon > 0$

$$\partial E_k \cap B_1 \subset \{|x_n| < \varepsilon\}$$

for all large enough k.

4.2. Phase transitions. Next we present density estimates for level sets of minimizers

$$J(u, \Omega) = \int_\Omega \frac{1}{2}|\nabla u|^2 + W(u)\, dx, \quad |u| \leq 1.$$

which were obtained by Cordoba and Caffarelli [**17**]. In this setting the domain Ω is large and the density estimates hold for large enough balls.

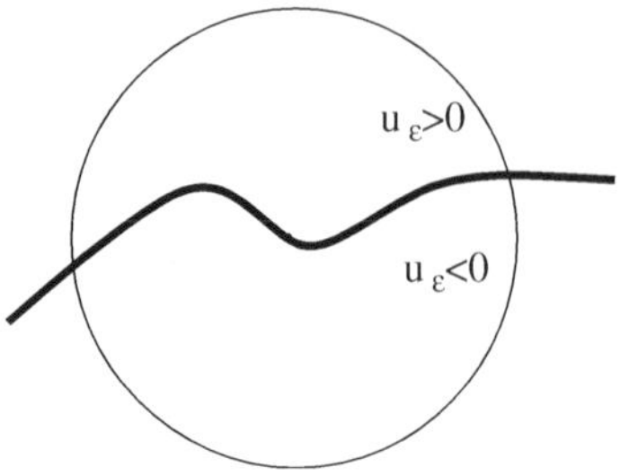

FIGURE 6. Density estimate for u_ε

Before we state the precise statement we first give a bound for the growth of $J(u, B_r)$.

PROPOSITION 4.3. *Assume* $r \geq 1$ *and* $B_{r+2} \subset \Omega$. *Then*

$$J(u, B_r) \leq C r^{n-1}, \tag{4.1}$$

where C *depends on* n *and* W.

PROOF. We compare u with the compact perturbation

$$v = \min\{u, h\} \quad \text{with } h(x) := \min\{(|x| - r)^+ - 1, 1\},$$

and obtain

$$J(u, B_r) \leq J(u, \{u < v\}) \leq J(h, \{u < v\}) \leq J(h, B_{r+2}) \leq C r^{n-1}.$$

□

THEOREM 4.4 (Caffarelli-Cordoba). *Given* $\alpha > -1$, $\beta < 1$, *if* u *is a minimizer of* J *in* B_R *and* $u(0) \geq \alpha$, *then*

$$|\{u > \beta\} \cap B_r| \geq c r^n$$

for $r \geq r_0(\alpha, \beta)$, *where* c *is a constant depending on* n *and* W.

PROOF. The proof consists of two steps. In the first step we prove a weaker version when α and β are close to -1 and in the second step we extend it to the general case. The ideas are similar to the minimal surface case but the proof is more involved. Let λ_0 be close to -1 so that W is increasing in the interval $[-1, \lambda_0]$.

Step 1. Let $-1 < \lambda \leq \lambda_0$. We prove the theorem with $\alpha = 1 + 2\lambda$, $\beta = \lambda$ and with the constant c depending also on λ.

For each $r \geq 1$ denote by

$$V(r) := |\{u > \lambda\} \cap B_r|, \quad a(r) := \int_{B_r} W(u)\, dx.$$

We want to deduce a discrete differential inequality involving V and a that resembles the one for minimal surfaces. We use as comparison function in B_r the radially symmetric function

$$v := -1 + 2e^{|x|-r}.$$

The test function v was chosen such that $v = 1$ on ∂B_r and satisfies

$$|\nabla v|^2 \leq C(W(v) + \chi_{\{v>0\}}) \leq C(W(v) + \chi_{\{v>\lambda\}}),$$

for some constant C depending on W.

Denote by K the closure of the open set $\{u > v\}$, and clearly

$$J(u, K) \leq J(v, K).$$

As in Section 2, the coarea formula gives

$$\int_{-1}^{1} \mathcal{H}^{n-1}(\{w = s\} \cap K)\sqrt{2W(s)}ds \leq J(w, K),$$

for any function w with values in $[-1, 1]$.

For each $s \in (-1, 1)$ denote (see Figure 7)

$$E_s := \{v < s < u\}.$$

We have

$$\partial E_s = (\{u = s\} \cup \{v = s\}) \cap K$$

and by the isoperimetric inequality

$$|E_s|^{\frac{n-1}{n}} \leq \mathcal{H}^{n-1}(\{u = s\} \cap K) + \mathcal{H}^{n-1}(\{v = s\} \cap K).$$

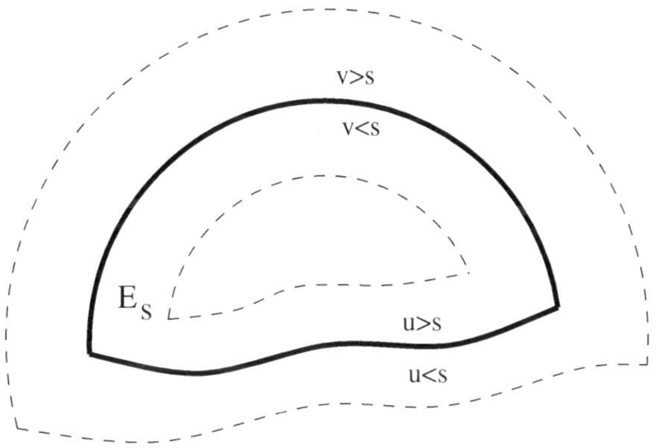

FIGURE 7. The set E_s

Multiplying with $\sqrt{2W(s)}$ and integrating in s we obtain

$$\int_{-1}^{1} |E_s|^{\frac{n-1}{n}} \sqrt{2W(s)}\, ds \leq J(u,K) + J(v,K) \leq 2J(v,K).$$

When $s < \lambda$, we have the inclusion $\{u > \lambda\} \cap B_r \subset E_s$ thus

$$V(r) = |\{u > \lambda\} \cap B_r| \leq |E_s|.$$

The above integral inequality implies

$$c(\lambda) V(r)^{\frac{n-1}{n}} \leq J(v,K),$$

with $c(\lambda) > 0$ a small constant that depends also on λ.

Also $J(u,K) \leq J(v,K)$ gives

$$a(r) \leq J(v,K),$$

hence

$$c(\lambda)(a(r) + V(r)^{\frac{n-1}{n}}) \leq J(v,K).$$

Next we estimate the right hand side $J(v,K)$ by splitting K into two domains

$$K_1 := B_{r-T} \cap K, \quad K_2 := K \cap (B_r \setminus B_{r-T}),$$

where T is a large constant to be made precise later. Clearly,

$$J(v,K_1) \leq J(v, B_{r-T}) \leq Ce^{-T} r^{n-1}.$$

For the second integral we use the inequality on $|\nabla v|^2$ together with $v \leq u$ on K and W increasing on $[0,\lambda]$:

$$\begin{aligned} J(v,K_2) &\leq C \int_{K_2} W(v) + \chi_{\{v>\lambda\}}\, dx \\ &\leq C \int_{K_2} W(u) + \chi_{\{u>\lambda\}}\, dx \\ &\leq C(V(r) - V(r-T) + a(r) - a(r-T)). \end{aligned}$$

In conclusion

$$c(\lambda)(a(r) + V(r)^{\frac{n-1}{n}}) \leq V(r) + a(r) - (V(r-T) + a(r-T)) + e^{-T} r^{n-1}.$$

From the Euler-Lagrange equation for the minimizer we find that u is Lipschitz in B_1 with the Lipschitz norm depending only on W. Since

$$u(0) \geq \alpha = 1 + 2\lambda$$

we find that $V(1) \geq \mu_0(\lambda)$. This and the inequality above easily implies that the sequence

$$d_k := a(kT) + V(kT)$$

satisfies a discrete differential inequality

$$c(\lambda)d_k^{\frac{n-1}{n}} \leq d_{k+1} - d_k + e^{-T}T^{n-1}k^{n-1}, \quad d_1 \geq \mu_0(\lambda).$$

The coefficient of k^{n-1} tends to 0 as $T \to \infty$. It is straightforward to check that

$$d_k \geq \mu_1 k^n$$

for some small $\mu_1(\lambda)$ provided that T is chosen large depending on λ. This gives

$$V(r) + a(r) \geq c(\lambda)r^n.$$

The energy bound (4.1) implies $a(r) \leq Cr^{n-1}$, thus

$$V(r) \geq c(\lambda)r^n \quad \text{if } r \geq 1,$$

and Step 1 is proved.

Step 2. We remove the assumptions on α, β and c by increasing the initial radius $r \geq r_1(\alpha, \beta)$.

Since $u(0) \geq \alpha$ we can find a small λ depending on α with $-1 < \lambda \leq \lambda_0$, and $\lambda \leq (\alpha - 1)/2$. By Step 1 there exists $c(\alpha) > 0$ such that for all $r \geq 1$

$$|\{u > \lambda\} \cap B_r| \geq c(\alpha)r^n.$$

On the other hand, the energy bound (4.1) gives,

$$|\{\beta \geq u > \lambda\} \cap B_r| \leq C(\lambda, \beta)J(u, B_r) \leq C(\lambda, \beta)r^{n-1},$$

hence

$$|\{u > \beta\} \cap B_r| \geq c(\alpha)r^n \tag{4.2}$$

for all $r > r_0(\alpha, \beta)$.

Finally we remark that in (4.2) we can take the constant to be independent of α by possibly increasing $r_0(\alpha, \beta)$.

Indeed, by (4.2) we know that there exists a constant $C(\alpha) > 0$ and a point $x_0 \in B_{C(\alpha)}$ such that $u(x_0) \geq 0$. Now we can apply (4.2) with $\alpha = 0$ by taking x_0 to be the origin, and obtain

$$|\{u > \beta\} \cap B_r(x_0)| \geq cr^n$$

for all $r \geq r_0(\beta)$, where c depends only on W. In conclusion

$$|\{u > \beta\} \cap B_r| \geq cr^n$$

for all $r > r_1(\alpha, \beta)$ and the theorem is proved. □

4.3. Asymptotic behavior. As a consequence of the density estimates we obtain that the level sets of u are asymptotically flat at ∞ at least in low dimensions. If $u : \mathbb{R}^n \to [-1, 1]$ is a global minimizer then, by Modica's theorem, the rescalings

$$u_{\varepsilon_k}(x) = u(x/\varepsilon_k)$$

satisfy

$$u_{\varepsilon_k} \to \chi_E - \chi_{\mathcal{C}E} \quad \text{in } L^1_{loc}(\mathbb{R}^n),$$

with ∂E a global minimal surface. Then the density estimates imply, as in the minimal surface case, that the level sets

$$\{u_{\varepsilon_k} = 0\} = \varepsilon_k \{u = 0\}$$

converge uniformly on compact sets to ∂E. Since ∂E is a global minimal surface, then ∂E is a hyperplane if $n \leq 7$, say

$$\partial E = \{x_n = 0\}.$$

Then $\{u = 0\}$ is asymptotically flat at ∞, i.e. there exist sequences θ_k, l_k with $l_k \to \infty$, $\theta_k / l_k \to 0$ such that

$$\{u = 0\} \cap B_{l_k} \subset \{|x_n| \leq \theta_k\}.$$

5. Flatness theorem of De Giorgi

In next two sections we give a new proof of the flatness theorem of De Giorgi that we mentioned in the introduction. The methods seem to be quite general and they can be applied to other equations or variational problems. Later we will use the same ideas to prove the corresponding flatness theorem for phase transitions.

The precise statement of the flatness theorem is the following.

THEOREM 5.1 (De Giorgi). *Assume E is minimal in B_1, $0 \in \partial E$ and*

$$\partial E \cap B_1 \subset \{|x_n| \leq \varepsilon_0\}$$

with $\varepsilon_0(n)$ small depending only on n. Then ∂E is an analytic surface in $B_{1/2}$.

The difficulty of this theorem lies in the fact that ∂E cannot be written as a graph. De Giorgi's proof relies on the monotonicity formula for minimal surfaces and the approximation of ∂E by harmonic functions (see [**40**]). We will give a proof based on Harnack inequality and viscosity solutions methods. First we give a different version of the theorem above known as improvement of flatness.

THEOREM 5.2 (Improvement of flatness). *Assume E is minimal in B_1, $0 \in \partial E$ and*

$$\partial E \cap B_1 \subset \{|x_n| \leq \varepsilon\}$$

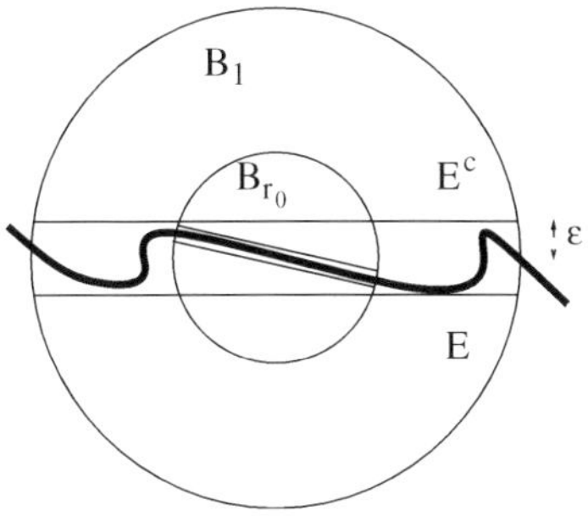

FIGURE 8. Improvement of flatness

with $\varepsilon \leq \varepsilon_0(n)$. *Then there exists a unit vector* ν_1 *such that*

$$\partial E \cap B_{r_0} \subset \left\{ |x \cdot \nu_1| \leq \frac{\varepsilon}{2} r_0 \right\},$$

where r_0 *is a small universal constant.*

This theorem implies that ∂E is a $C^{1,\alpha}$ graph in $B_{3/4}$. To see this we apply the theorem inductively and we obtain unit vectors ν_k such that

$$\partial E \cap B_{r_0^k} \subset \left\{ |x \cdot \nu_k| \leq \frac{\varepsilon}{2^k} r_0^k \right\}.$$

This gives

$$|\nu_{k+1} - \nu_k| \leq C(r_0) \frac{\varepsilon}{2^k}$$

hence $\nu_k \to \nu(0)$ and moreover

$$|\nu_k - \nu(0)| \leq C \frac{\varepsilon}{2^k}.$$

We obtain

$$\partial E \cap B_{r_0^k} \subset \left\{ x \cdot \nu(0) \leq C \frac{\varepsilon}{2^k} r_0^k = C \varepsilon r_0^{k(1+\alpha)} \right\},$$

which implies that ∂E is a differentiable surface at 0 with normal $\nu(0)$. Applying this argument at all points in $\partial E \cap B_{3/4}$ we see that ∂E is in fact a $C^{1,\alpha}$ surface.

Once the surface is a $C^{1,\alpha}$ graph, we apply Schauder's estimates (see [**39**]) for the minimal surface equation satisfied by the graph and obtain that in $B_{1/2}$, ∂E is smooth and in fact analytic.

We will derive the improvement of flatness theorem from the following weaker result.

THEOREM 5.3 (Harnack inequality). *Assume* E *is minimal in* B_1 *and*

$$\partial E \cap B_1 \subset \{|x_n| \leq \varepsilon\}$$

with $\varepsilon \leq \varepsilon_1(n)$. *Then*

$$\partial E \cap B_{1/2} \subset \{|x_n| \leq \varepsilon(1 - \eta)\},$$

where $\eta > 0$ *is a small universal constant.*

Before we proceed we need to introduce the notion of viscosity solutions for boundaries of sets.

DEFINITION. We say that the boundary of a set ∂F satisfies the mean curvature equation

$$M(D^2 v, \nabla v) := (1 + |\nabla v|^2)\triangle v - (\nabla v)^T D^2 v \, \nabla v = 0$$

in the viscosity sense if for any smooth function φ which has the subgraph

$$S := \{x_n < \varphi(x')\}$$

included in F or $\mathcal{C}F$ in a small ball $B_\rho(Y)$ around some point

$$Y = (y', y_n) \in \partial F \cap \partial S$$

we have

$$M(D^2\varphi, \nabla\varphi) \leq 0.$$

If we consider the supergraph of φ then the opposite inequality holds.

LEMMA 5.4. *If ∂E is a minimal surface then ∂E satisfies the mean curvature equation in the viscosity sense.*

A proof of this lemma can be found in [**16**]. The idea is that if E contains the subgraph of φ and $M(D^2\varphi, \nabla\varphi) > 0$, then we can decrease the perimeter by adding to E the small domain

$$\{\varphi(x') < x_n < \varphi(x') + \varepsilon^2 - \varepsilon|x'|^2\}.$$

Next we explain why Harnack inequality gives the improvement of flatness theorem. The main reason is that Harnack inequality allows us to approximate a flat minimal surface by a harmonic function.

Theorem 5.3 implies Theorem 5.2

The proof is by compactness. Assume by contradiction the statement of Theorem 5.2 is not true. Then we can find a sequence of minimal surfaces ∂E_k which satisfy the hypothesis

$$0 \in \partial E_k \cap B_1 \subset \{|x_n| < \varepsilon_k\}$$

with $\varepsilon_k \to 0$ for which the conclusion does not hold.

At each point $x_0 \in \partial E_k \cap B_{1/2}$ we apply Harnack inequality in $B_{1/2}(x_0)$ and obtain that

$$\partial E_k \cap B_{1/4}(x_0) \subset \{|(x - x_0) \cdot e_n| \leq 2\varepsilon_k(1 - \eta)\}.$$

We apply Harnack inequality repeatedly as long as the hypothesis is satisfied. We obtain that for all $m \geq 2$ such that

$$\varepsilon_k 2^m (1 - \eta)^{m-2} < \varepsilon_1(n)$$

we have the inclusion

$$\partial E_k \cap B_{2^{-m}}(x_0) \subset \{|(x - x_0) \cdot e_n| \leq 2\varepsilon_k(1 - \eta)^{m-1}\}.$$

Clearly we can take $m \to \infty$ as $\varepsilon_k \to 0$.

We strech this picture by a factor ε_k^{-1} in the x_n direction. The sets

$$A_k := \left\{ (x', \frac{x_n}{\varepsilon_k}) | \, (x', x_n) \in \partial E_k \cap B_1 \right\}$$

are included in $\{|x_n| \leq 1\}$. Moreover, for each m as above

$$A_k \cap \left\{ |x' - x_0'| \leq 2^{1-m} \right\} \subset \left\{ |(x - x_0) \cdot e_n| \leq 2(1-\eta)^{m-1} \right\}.$$

This gives a Hölder modulus of continuity of A_k outside a $C\varepsilon_k^{-1}$ neighborhood of 0. By Arzela-Ascoli theorem we can assume that as $\varepsilon_k \to 0$, by passing if necessary to a subsequence, A_k converges in Hausdorff distance to the graph of a Hölder continuous function $(x', w(x'))$ in $\{|x'| \leq 1/2\}$.

Next we show that w is harmonic in the viscosity sense, that is w cannot be touched by below (above) at an interior point by a strict subharmonic (superharmonic) quadratic polynomial.

Assume $P(x')$ is a quadratic polynomial whose graph touches by below the graph of w at some point. Then A_k touches a translation of P which implies that ∂E_k touches the graph of $\varepsilon_k P(x') + c$ at some interior point. Clearly ∂E_k satisfies the minimal surface equation in the viscosity sense hence

$$M(\varepsilon_k D^2 P, \varepsilon_k \nabla P) = \varepsilon_k \triangle P + \varepsilon_k^3 (|\nabla P|^2 \Delta P - (\nabla P)^T \, D^2 P \, \nabla P) \leq 0$$

at the contact point. We let $\varepsilon_k \to 0$ and obtain

$$\triangle P \leq 0.$$

Since w is harmonic, $w(0) = 0$ (since $0 \in A_k$) and $|w| \leq 1$, we find that

$$|w(x') - x' \cdot \nabla w(0)| \leq \frac{r_0}{4} \quad \text{if } |x'| \leq 2r_0,$$

provided that r_0 is chosen small, universal. This easily implies

$$A_k \cap \left\{ |x'| \leq \frac{3}{2} r_0 \right\} \subset \left\{ |x_n - x' \cdot \nabla w(0)| \leq \frac{r_0}{3} \right\}$$

for large k hence,

$$\partial E_k \cap B_{r_0} \subset \left\{ |x \cdot \nu_k| \leq \frac{\varepsilon_k}{2} r_0 \right\} \quad \text{for some } |\nu_k| = 1.$$

Thus ∂E_k satisfies the conclusion of the Theorem and we reach a contradiction. $\square$

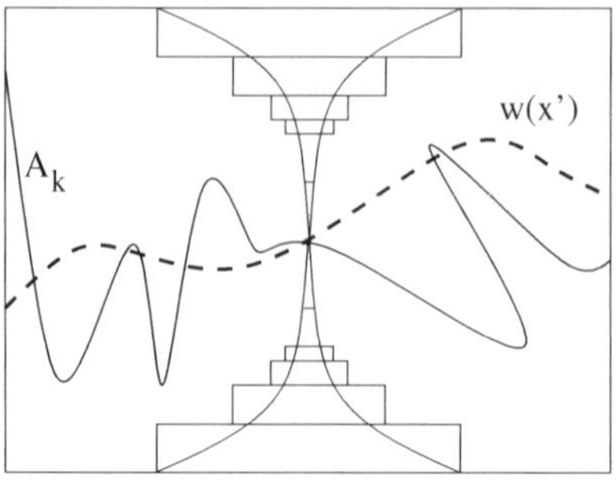

FIGURE 9. Convergence of A_k

6. Harnack inequality

In this section we will give a proof of Harnack inequality, Theorem 5.3. The ideas come from the proof of the classical Harnack inequality for uniformly elliptic second order linear equations due to Krylov and Safonov [**44**]. Let us briefly explain the key steps in their proof. First, they obtain an estimate in L^∞ from one point to a set of points of fixed positive measure. Then, they iterate this step using Calderon-Zygmund decomposition and obtain L^∞ estimates on a set of almost full measure, from which Harnack inequality can be deduced.

Of course these results do not apply directly in our case. The mean curvature equation $M(D^2 v, \nabla v) = 0$ is not uniformly elliptic since the ellipticity constants depend on $|\nabla v|$. Moreover, the set ∂E cannot be written as a graph of a function v. However, the measure estimates still hold. This is because the set on which L^∞ estimates hold is obtained as a contact set with a family of paraboloids that lie below the graph of v. Due to the flatness assumption, at the contact points the gradient of v is small, and the equation becomes uniformly elliptic. Also, the minimality and flatness of ∂E imply that from the measure point of view, ∂E is almost a graph. Below we give the details. A systematic treatment of estimates for flat solutions to general elliptic equations can be found in [**57**].

First we prove a measure estimates for a general elliptic equation in $\mathbb{R}^n$.

$$F(D^2 v, Dv) = 0, \quad v : B_1 \to \mathbb{R}, \tag{6.1}$$

with

$$F : \mathcal{S} \times \mathbb{R}^n \to \mathbb{R}$$

where $\mathcal{S}$ denotes the space of $n \times n$ symmetric matrices. We assume that F is uniformly elliptic only in a neighborhood of $(0, 0) \in \mathcal{S} \times \mathbb{R}^n$. Precisely, assume there exist $\delta > 0$ small and λ, Λ such that if $|p| \leq \delta$ and $N \geq 0$ is positive symmetric then

$$F(0,p) = 0,$$
$$\lambda|N| \leq F(M+N,p) - F(M,p) \leq \Lambda|N| \quad \text{if } |M|, |N| \leq \delta,$$
$$0 \leq F(M+N,p) - F(M,p) \quad \text{if } |M| < \delta.$$

Clearly, these hypotheses are satisfied by the mean curvature equation.

We prove our estimates for supersolutions v of the equation above. That is we assume v is continuous, and any quadratic polynomial P that touches v by below at an interior point x_0 satisfies

$$F(D^2P, DP(x_0)) \leq 0.$$

PROPOSITION 6.1 (Estimate in measure). *Let $v : B_1 \to \mathbb{R}$, $v \geq 0$ be a viscosity supersolution to equation (6.1). Given $\mu > 0$ small, there exist ε small and M large depending on μ and n,δ, λ, Λ such that if*

$$v(0) \leq \varepsilon,$$

then

$$|\{v > Mv(0)\} \cap B_{1/3}| < \mu.$$

In this section positive constants depending on n, λ, Λ and δ are called universal constants.

For a small number $a > 0$, we consider the paraboloids with center at $y \in \bar{B}_1$ and opening $-a$, i.e.

$$-\frac{a}{2}|x-y|^2 + const.$$

We slide them vertically from $-\infty$ till they touch the graph of v. We denote the set of interior contact points with D_a (see Figure 10). Precisely,

$$D_a := \Big\{ z \in B_1 \mid \exists y \in \bar{B}_1 \quad \text{s.t. } v(z) + \frac{a}{2}|z-y|^2 \leq v(x)$$
$$+ \frac{a}{2}|x-y|^2 \quad \forall x \in B_1 \Big\}.$$

We prove Proposition 6.1 by estimating the size of D_a. Notice that

$$v(z) \leq v(0) + 2a, \quad \forall z \in D_a.$$

First we obtain an estimate of the contact set of z's in terms of the set of centers y that we start. This is a variant of Alexandrov-Bakelman-Pucci estimate for uniformly elliptic equations.

LEMMA 6.2 (ABP estimate). *Let $a > 0$ be small, $a \leq \delta/2$, and let $F \subset \bar{B}_1$ be a closed set. For each $y \in F$, we slide the paraboloid of opening $-a$ and center y by below till we touch the graph of v for the first time.*

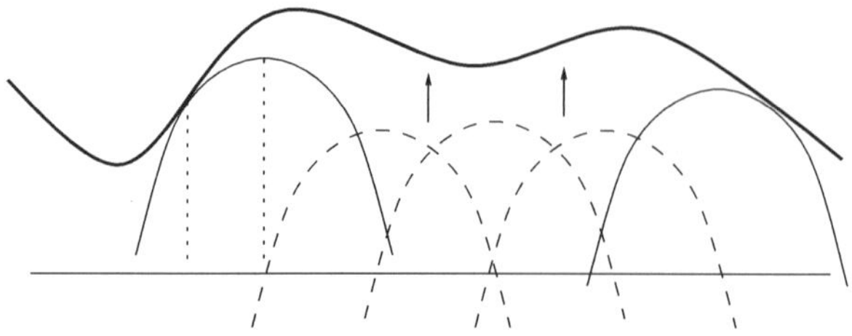

FIGURE 10. Sliding paraboloids

Suppose the set of all contact points, denoted by E, is included in a compact set of B_1. Then

$$|E| \geq c|F|,$$

with $c > 0$ a small universal constant.

PROOF. Let us assume for simplicity that v is smooth. Otherwise one needs to regularize v using a standard *inf*-convolution.

At a contact point z, the corresponding center $y(z)$ is given by

$$y(z) = z + \frac{1}{a} Dv(z).$$

Since $y, z \in \bar{B}_1$ we see that $|Dv(z)| \leq 2a \leq \delta$. The differential map is

$$D_z y = I + \frac{1}{a} D^2 v(z)$$

and moreover, since v has at z a tangent paraboloid of opening $-a$ by below,

$$D^2 v(z) \geq -aI.$$

Now we claim that the equation implies

$$D^2 v(z) \leq CaI,$$

with C a large constant depending on n, δ, λ, Λ. Indeed, otherwise

$$D^2 v(z) \geq Cae \otimes e - aI,$$

for some unit vector e hence

$$F(D^2 v(z), Dv(z)) \geq F(Cae \otimes e - aI, p) \geq \lambda(C-1)a - (n-1)\Lambda a > 0,$$

and we contradict that v is a supersolution.

In conclusion

$$0 \leq D_z y \leq CI,$$

which gives

$$|F| = \int_E |\det D_z y| \, dz \leq C|E|.$$

□

LEMMA 6.3. *There exist positive universal constants C, c such that if $a \le C^{-1}\delta$ and*

$$D_a \cap B_r(x_0) \neq \emptyset$$

for some ball $\bar{B}_r(x_0) \subset B_1$, then

$$|D_{Ca} \cap B_{r/8}(x_0)| \ge cr^n.$$

PROOF. Let

$$z_1 \in B_r(x_0) \cap D_a,$$

be a contact point and denote by $y_1 \in \bar{B}_1$ the center of the tangent paraboloid

$$P(x, y_1) = -\frac{a}{2}|x - y_1|^2 + v(z_1) + \frac{a}{2}|z_1 - y_1|^2,$$

that touches v by below at z_1. First we find a point $x_1 \in \bar{B}_{r/16}(x_0)$ such that

$$v(x_1) - P(x_1, y_1) \le Car^2, \quad C \text{ universal.} \tag{6.2}$$

Let $\phi : \bar{B}_1 \to \mathbb{R}^+$ be the radially symmetric continuous function

$$\phi(x) = \begin{cases} \alpha^{-1}(|x|^{-\alpha} - 1), & 1/16 \le |x| \le 1 \\ \alpha^{-1}(16^\alpha - 1), & |x| < 1/16. \end{cases}$$

where α is a large universal constant. Construct a function ψ (see Figure 11) by adding a rescaling of the above function to the tangent paraboloid $P(x, y_1)$ i.e.

$$\psi(x) := P(x, y_1) + ar^2\phi\left(\frac{x - x_0}{r}\right).$$

We claim that ψ is a strict subsolution of our equation in the annular region

$$\frac{r}{16} < |x - x_0| < r.$$

Indeed,

$$|D\psi| \le |DP| + |D\phi| \le Ca \le \delta,$$

and, by choosing α sufficiently large,

$$\begin{aligned} F(D^2\psi, D\psi) &= F(-aI + aD^2\phi, p) \\ &\ge \lambda((\alpha+1)t^{-\alpha-2} - 1) - \Lambda(n-1)(1 + t^{-\alpha-2}) > 0 \end{aligned}$$

where

$$t =: \frac{|x - x_0|}{r}, \quad \frac{1}{16} \le t \le 1.$$

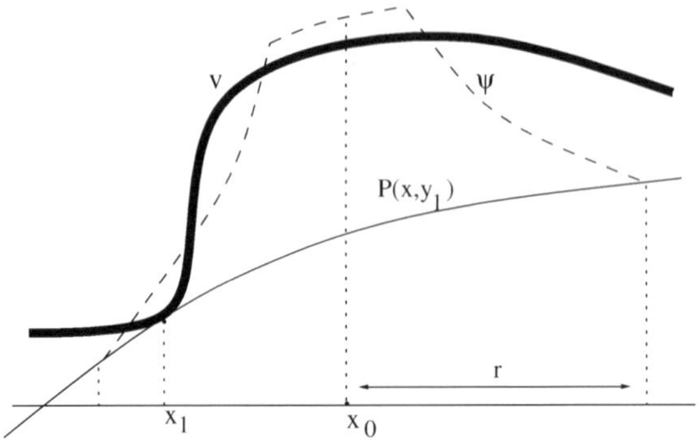

FIGURE 11. The function ψ

Now we slide the graph of ψ from below till we touch the graph of v for the first time. In other words we look for the point x_1 where the minimum

$$\min_{\bar{B}_r(x_0)} (v - \psi),$$

is realized. Notice that the minimum value is negative since

$$v(z_1) - \psi(z_1) = P(z_1, y_1) - \psi(z_1) = -ar^2\phi\left(\frac{z_1 - x_0}{r}\right) < 0.$$

This implies that x_1 does not belong to $\partial B_r(x_0)$ because

$$v(x) - \psi(x) \geq P(x, y_1) - P(x, y_1) = 0 \quad \text{if } x \in \partial B_r(x_0).$$

Since ϕ is a strict subsolution in the annular region above we conclude that $x_1 \in B_{r/16}(x_0)$ and

$$v(x_1) < \psi(x_1) \leq P(x_1, y_1) + Car^2$$

which proves (6.2).

Now we slide from below the family of paraboloids

$$P(x, y_1) - C'\frac{a}{2}|x - y|^2 + c_y, \quad \text{with } |y - x_1| \leq r/64,$$

till they become tangent to the graph of v. The opening of the above paraboloid is $-(C' + 1)a$ and the center

$$\frac{C'}{C' + 1}y + \frac{1}{C' + 1}y_1.$$

The centers range over a ball of radius cr as $y \in B_{r/64}(x_1)$. From (6.2) we find

$$c_y \leq Car^2 + C'\frac{a}{2}(r/64)^2.$$

If $|x - x_1| \geq r/16$ and C' is sufficiently large,

$$P(x, y_1) - C'\frac{a}{2}|x - y|^2 + c_y \leq P(x, y_1) - C'\frac{a}{2}(r/32)^2 + Car^2 + C'\frac{a}{2}(r/64)^2$$
$$< P(x, y_1) \leq v(x).$$

Thus, the contact points are inside $B_{r/16}(x_1) \subset B_{r/8}(x_0)$. From the previous lemma we obtain

$$|D_{(C'+1)a} \cap B_{r/8}(x_0)| \geq cr^n,$$

with c small universal, and the lemma is proved. □

Next we prove a simple measure covering lemma.

LEMMA 6.4 (Covering lemma). *Assume the closed sets F_k satisfy*

$$F_0 \subset F_1 \subset F_2 \cdots \subset \bar{B}_{1/3}, \quad F_0 \neq \emptyset$$

and for any x, r such that

$$B_{r/8}(x) \subset B_{1/3}, \quad \bar{B}_r(x) \subset B_1,$$
$$F_k \cap B_r(x) \neq \emptyset,$$

then

$$|F_{k+1} \cap B_{r/8}(x)| \geq cr^n.$$

Then

$$|B_{1/3} \setminus F_k| \leq (1 - c_1)^k |B_{1/3}|$$

for some small constant c_1 depending on c.

PROOF. Let $x_0 \in B_{1/3} \setminus F_k$ and let

$$r := dist(x_0, F_k).$$

We first prove that

$$|F_{k+1} \cap B_{r/3}(x_0)| \geq c_1 |B_r(x_0) \cap B_{1/3}|. \tag{6.3}$$

Let (see Figure 12)

$$x_1 := x_0 - \frac{r}{6}\frac{x_0}{|x_0|},$$

and it is easy to check that

$$B_{r/6}(x_1) \subset B_{r/3}(x_0) \cap B_{1/3}, \quad \bar{B}_{7r/6}(x_1) \subset B_1.$$

Since

$$dist(x_1, F_k) \leq r + \frac{r}{6},$$

we apply the hypothesis and conclude

$$|F_{k+1} \cap B_{r/6}(x_1)| \geq cr^n \geq c_1 |B_r(x_0) \cap B_{1/3}|,$$

which proves (6.3).

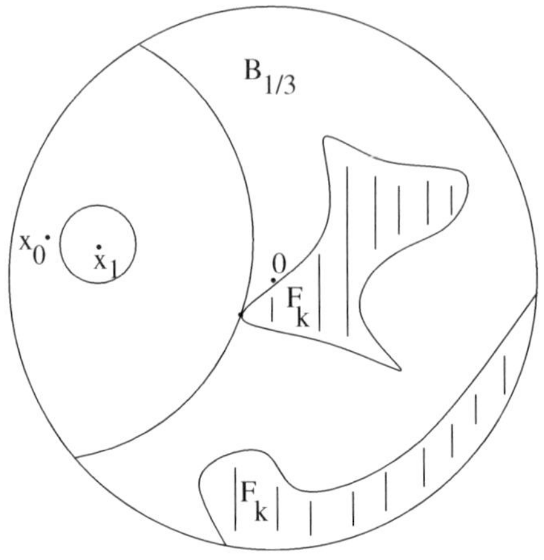

FIGURE 12. Covering of $\mathcal{C}F_k$

For each $x \in B_{1/3} \setminus F_k$ we let $r = dist(x, F_k)$. From the family $B_r(x)$ we choose a Vitali subcover, i.e. balls $B_{r_i}(x_i)$ that cover $B_{1/3} \setminus D_k$ for which $B_{r_i/3}(x_i)$ are disjoint.

We have

$$\begin{aligned}|B_{1/3} \setminus F_k| &\leq \sum |B_{r_i}(x_i) \cap B_{1/3}| \\ &\leq \sum c_1^{-1} |B_{r_i/3}(x_i) \cap F_{k+1}| \leq c_1^{-1} |F_{k+1} \setminus F_k|,\end{aligned}$$

which implies

$$|B_{1/3} \setminus F_{k+1}| \leq |B_{1/3} \setminus F_k| - |F_{k+1} \setminus F_k| \leq (1 - c_1)|B_{1/3} \setminus F_k|.$$

□

PROOF OF PROPOSITION 6.1. Let

$$a = 20v(0),$$

and define

$$F_k := D_{C^k a} \cap \bar{B}_{1/3},$$

where C is the constant from Lemma 6.3. Since $v \geq 0$ in B_1, the paraboloid of opening $-a$ and center 0 touches the graph of v for the first time in $\bar{B}_{1/3}$, hence $F_0 \neq \emptyset$. Moreover, by Lemma 6.3 the hypothesis of the Covering lemma are satisfied as long as $C^k a \leq \delta$. Thus

$$|B_{1/3} \setminus D_{C^k a}| \leq (1 - c_1)^k |B_{1/3}|, \quad \text{if } C^k a \leq \delta.$$

In the set $D_{C^k a}$ the function v is less than $3C^k a$, and we prove the proposition by first choosing k large depending on μ and then $v(0) \leq \varepsilon$ small, so that $20C^k \varepsilon \leq \delta$. □

Finally we can give the proof of Theorem 5.3 by applying the results above to the mean curvature equation in $\mathbb{R}^{n-1}$.

PROOF OF THEOREM 5.3. Let ∂E be a minimal surface in B_1 with

$$0 \in \partial E \subset \{|x_n| \le \varepsilon\}.$$

Assume without loss of generality that in B_1 the set E contains $\{x_n < -\varepsilon\}$ and is contained in $\{x_n \le \varepsilon\}$. We view ∂E as a multivalued graph over the $x' \in \mathbb{R}^{n-1}$ that satisfies the mean curvature equation in the viscosity sense. We denote by $B'_r(x')$ the $n-1$ dimensional ball of center x' and radius r and by π_n the projection along e_n in $\mathbb{R}^{n-1}$.

It suffices to prove that

$$\partial E \cap B_{1/10} \subset \{x_n \ge -\varepsilon(1-\eta)\}$$

for some small constant $\eta > 0$ depending on n. Assume by contradiction that there exists

$$z \in \partial E \cap B_{1/10}, \quad \text{with } z_n \le -\varepsilon(1-\eta).$$

Then we can use the argument in Proposition 6.1 to show the existence of a subset D of contact points (see Figure 9)

$$D \subset \partial E \cap \{|x' - z'| < 3/10, \quad x_n \le -\varepsilon + M\eta\varepsilon\}$$

with

$$\mathcal{H}^{n-1}(\pi_n(D)) \ge (1-\mu)\mathcal{H}^{n-1}(B'_{3/10}(z')),$$

with μ a small universal constant to be made precise later. We choose η sufficiently small depending on μ such that $M\eta < 1/2$. Then D lies below the plane $x_n = -\varepsilon/2$.

Next we use Lemma 6.2 to find another closed subset of contact points in ∂E (see Figure 13),

$$A \subset \partial E \cap B_{1/10} \cap \{x_n \ge -\varepsilon/2\}$$

so that

$$\mathcal{H}^{n-1}(\pi_n(A)) > c_0$$

with $c_0 > 0$ a small constant depending only on n. For this we slide paraboloids

$$x_n = \frac{a}{2}|x' - y'|^2 + const$$

of opening $a = 10^3\varepsilon$ and centers $|y'| \le 10^{-2}$ by above till they touch ∂E. Since $0 \in \partial E$, it is easy to check that the contact set A is included in the set

$$\{|x'| < 1/10, \quad -\varepsilon/2 \le x_n \le \varepsilon\}.$$

The bound on $\mathcal{H}^{n-1}(\pi_n(A))$ follows from the measure estimate of Lemma 6.2 because the family of centers y' projects along e_n into a set of fixed positive measure.

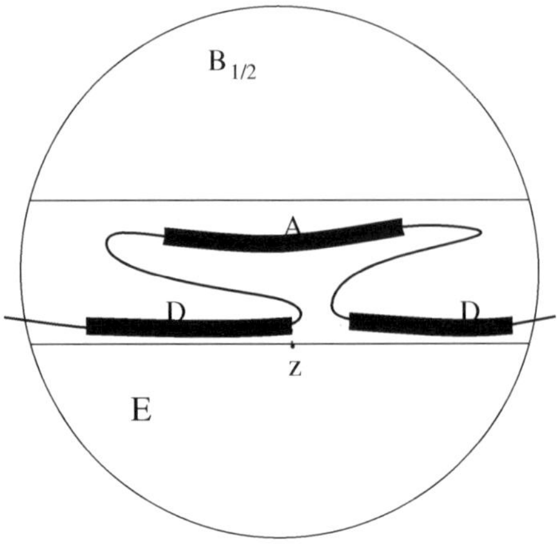

FIGURE 13. The contact sets A and D

The sets A and D are disjoint since they lie on different sides of the plane $x_n = -\varepsilon/2$. Moreover, they both project along e_n in $B'_{3/10}(z')$. We choose μ small universal such that

$$\mathcal{H}^{n-1}(\pi_n(A)) + \mathcal{H}^{n-1}(\pi_n(D)) \geq \mathcal{H}^{n-1}(B'_{3/10}(z')) + c_0/2.$$

This implies that the projections $\pi_n(A)$ and $\pi_n(D)$ have an intersection of at least $c_0/2$ measure. This gives

$$P_{B_{1/2}}(E) \geq \mathcal{H}^{n-1}(B'_{1/2}) + c_0/3.$$

On the other hand, by comparing E with $\{x_n \leq -\varepsilon\}$ in $B_{1/2}$, we find from the minimality of E that

$$P_{B_{1/2}}E \leq \mathcal{H}^{n-1}(B'_{1/2}) + C\varepsilon,$$

and we reach a contradiction if ε is small. □

7. The flatness theorem for minimizers

In the next three sections we prove the analogue of Theorem 5.2 for phase transitions, i.e. minimizers of the energy

$$J(u, \Omega) = \int_\Omega \frac{1}{2}|\nabla u|^2 + W(u)\,dx.$$

The corresponding flatness theorem is the following.

THEOREM 7.1 (Improvement of flatness). *Let u be a minimizer of J in the cylinder*

$$\{|x'| < l\} \times \{|x_n| < l\},$$

and assume that

$$0 \in \{u = 0\} \cap \{|x'| < l\} \subset \{|x_n| < \theta\}.$$

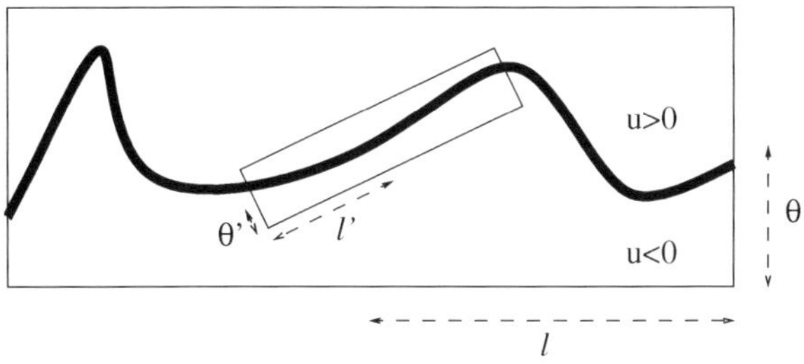

FIGURE 14. Improvement of flatness

Given $\theta_0 > 0$ there exists $\varepsilon_1(\theta_0) > 0$ depending on n, W and θ_0 such that if

$$\frac{\theta}{l} \leq \varepsilon_1(\theta_0), \quad \theta_0 \leq \theta$$

then

$$\{u = 0\} \cap \{|\pi_\xi x| < \eta_2 l\} \subset \{|x \cdot \xi| < \eta_1 \theta\},$$

for some unit vector ξ, where $0 < \eta_1 < \eta_2 < 1$ are constants depending only on n. (π_ξ denotes the projection along ξ)

The theorem is valid for any level set s with $|s| < 1$, but it was stated for $s = 0$ for simplicity. As a consequence of this flatness theorem we obtain the following important corollary:

COROLLARY 7.2. *Let u be a global minimizer of J in $\mathbb{R}^n$. Suppose that the 0 level set $\{u = 0\}$ is asymptotically flat at ∞, i.e. there exist sequences of positive numbers θ_k, l_k and unit vectors ξ_k with $l_k \to \infty$, $\theta_k l_k^{-1} \to 0$ such that*

$$\{u = 0\} \cap B_{l_k} \subset \{|x \cdot \xi_k| < \theta_k\}.$$

Then the 0 level set is a hyperplane and u is one-dimensional.

By saying that u is one-dimensional we understand that u depends only on one direction ξ, i.e. $u = g(x \cdot \xi)$ for some function g. In our case, g is a minimizer in 1D and can be computed explicitly from W. The function g is unique up to translations.

PROOF. Without loss of generality assume $u(0) = 0$. Fix $\theta_0 > 0$, and choose k large such that $\theta_k l_k^{-1} \leq \varepsilon \leq \varepsilon_1(\theta_0)$.

If $\theta_k \geq \theta_0$ then we apply Theorem 7.1 and obtain that $\{u = 0\}$ is trapped in a flatter cylinder. We apply Theorem 7.1 repeatedly till the height of the cylinder becomes less than θ_0.

In some system of coordinates we obtain

$$\{u = 0\} \cap \left(\{|y'| < l'_k\} \times \{|y_n| < l'_k\}\right) \subset \{|y_n| \leq \theta'_k\},$$

with $\theta_0 \geq \theta'_k \geq \eta_1 \theta_0$ and $\theta'_k l'^{-1}_k \leq \theta_k l_k^{-1} \leq \varepsilon$, hence

$$l'_k \geq \frac{\eta_1}{\varepsilon} \theta_0.$$

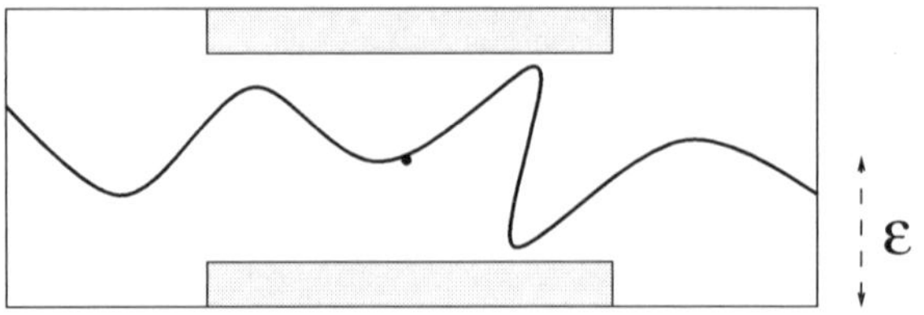

FIGURE 15. Harnack inequality

We let $\varepsilon \to 0$ and obtain that $\{u = 0\}$ is included in an infinite strip of width θ_0. We then let $\theta \to 0$ and obtain that $\{u = 0\}$ is a hyperplane. Similarly, all the level sets are hyperplanes which implies that u is one-dimensional. □

This corollary gives one of our main results: the uniqueness up to rotations of global minimizers in low dimensions (see Section 10).

The proof of Theorem 7.1 is quite involved. The idea is to reproduce the arguments from the minimal surface case given in the previous two sections. We present the proof of Theorem 7.1, with the exception of a number of technical lemmas that can be found in the original paper [**55**].

Again we will prove the improvement of flatness from a version of Harnack inequality.

THEOREM 7.3 (Harnack inequality for minimizers). *Let u be a minimizer of J in the cylinder*

$$\{|x'| < l\} \times \{|x_n| < l\}$$

and assume that

$$0 \in \{u = 0\} \subset \{|x_n| < \theta\}.$$

Given $\theta_0 > 0$ there exists $\varepsilon_0(\theta_0) > 0$ depending on n, W and θ_0, such that if

$$\theta l^{-1} \leq \varepsilon_0(\theta_0), \quad \theta_0 \leq \theta,$$

then

$$\{u = 0\} \cap \{|x'| < \eta_0 l\} \subset \{|x_n| < (1 - \eta_0)\theta\},$$

where $\eta_0 > 0$ is a small constant depending on n and W.

In the remaining part of this section we show that Theorem 7.3 implies Theorem 7.1. We start by introducing some notation.

Notation. We denote points as follows

$$X = (x, x_{n+1}) = (x', x_n, x_{n+1}) \in \mathbb{R}^{n+1}$$
$$X \in \mathbb{R}^{n+1}, \quad x' \in \mathbb{R}^{n-1}, \quad x \in \mathbb{R}^n, \quad |x_{n+1}| < 1.$$

We use the following notation:

- $B(x, r)$ is the ball of center x and radius r in $\mathbb{R}^n$;
- $\mathcal{B}(X, r)$ is the ball of center X and radius r in $\mathbb{R}^{n+1}$;
- $\angle(\nu_1, \nu_2) \in (0, \pi)$ is the angle between the vectors ν_1 and ν_2;
- $\pi_\nu X = X - (X \cdot \nu)\nu$ is the projection along ν;
- P_ν is the hyperplane perpendicular to ν going through the origin;
- $\pi_i := \pi_{e_i}$, $P_i := P_{e_i}$.

Constants depending on n, and W are called universal and we denote them by C, c, $\bar{C}_i$, $\bar{c}_i$. We write $\bar{C}_i$, $\bar{c}_i$ for constants that we use throughout the paper and by C, c various constants used in proofs that may change from line to line.

In the proof we often construct radial barriers

$$x_{n+1} = g(|x| - r),$$

from an increasing real function $g : I \to [-1, 1]$ from some interval I containing 0. In order to check wether or not such a barrier is a subsolution or supersolution we have to check inequalities involving g, g' and g''. In this cases we consider $s = g$ as the new variable and we define a new function on I

$$h(s) := \frac{1}{2}(\frac{dg}{dt})^2.$$

Straightforward computations give

$$g' = \sqrt{2h}, \quad g'' = h'$$

and now the inequalities involve only h and h' in I. We can reconstruct g from h (up to a translation) since

$$H(s) := \int_0^s \frac{1}{\sqrt{2h(\zeta)}} d\zeta = g^{-1}(s) - g^{-1}(0).$$

In particular, if we define

$$h_0(s) := W(s),$$
$$H_0(s) := \int_0^s \frac{1}{\sqrt{2h_0(\zeta)}} d\zeta,$$

and

$$g_0(t) := H_0^{-1}(t)$$

then we find

$$g_0''(t) = h_0'(g_0(t)) = W'(g_0),$$

thus, g_0 is the one dimensional solution of our semilinear equation.

First we show that the 0 level set of a minimizer satisfies in some sense an equation in the viscosity sense.

PROPOSITION 7.4. *Let u be a minimizer of J in $\{|x'| < l\} \times \{|x_n| < l\}$ and assume that $u(0) = 0$ and $u < 0$ below the graph of the quadratic polynomial*

$$\Gamma_1 := \left\{ x_n = P_1(x') = \frac{\theta}{l^2}\frac{1}{2}x'^T M_1 x' + \frac{\theta}{l}\xi \cdot x' \right\},$$

where M_1 is a $(n-1) \times (n-1)$ symmetric matrix. Also assume

$$\|M_1\| < \delta^{-1}, \quad |\xi| < \delta^{-1}$$

for some small $\delta > 0$. There exists $\sigma(\delta) > 0$ small, such that if

$$\theta l^{-1} \leq \sigma(\delta), \quad \theta \geq \delta,$$

then

$$tr\ M_1 \leq \delta.$$

As a consequence of this proposition we end up with a version of the Theorem of Modica in the viscosity sense. This can be viewed as the analogue of Lemma 5.4 to the case of minimizers of J.

PROPOSITION 7.5 (Modica theorem - viscosity sense). *Let u be a minimizer of J and assume that $u(0) = 0$. Consider the graph of a quadratic polynomial*

$$\Gamma = \left\{ x_n = P(x') := \frac{1}{2}x'^T M x' \right\},$$

that satisfies

$$\Delta P = tr\ M > \delta_0 \|M\|, \quad \|M\| \leq \delta_0^{-1}, \tag{7.1}$$

for some $\delta_0 > 0$ small.

Then Γ cannot touch by below $\{u_\varepsilon = 0\}$ at 0 in a $\delta_0(\Delta P)^{-\frac{1}{2}}\varepsilon^{\frac{1}{2}}$ neighborhood, if $\varepsilon \leq \sigma_0(\delta_0)$ where $\sigma_0(\delta_0) > 0$ depends on δ_0 and the universal constants.

By "Γ touches by below $\{u_\varepsilon = 0\}$ at 0 in a $\delta_0(\Delta P)^{-\frac{1}{2}}\varepsilon^{\frac{1}{2}}$ neighborhood" we understand

$$\{u_\varepsilon = 0\} \cap \{x_n < P(x')\} \cap \{|x| < \delta_0(\Delta P)^{-\frac{1}{2}}\varepsilon^{\frac{1}{2}}\} = \emptyset.$$

Proposition 7.5 says that $\{u_\varepsilon = 0\}$ satisfies the mean curvature equation in some weak viscosity sense in which we have to specify the size of the neighborhood around the touching point. The size of the neighborhood depends on both the polynomial P and ε.

If P is fixed and $\varepsilon \to 0$ then the radius of the neighborhood converges to 0. In particular, if $\{u_\varepsilon = 0\}$ converges uniformly to a surface, then this surface satisfies in the viscosity sense the mean curvature equation.

One way to interpret the above proposition is the following:
Suppose that P has positive mean curvature and let δ_0 be small such that (7.1) holds. In order to obtain a contradiction that P touches $\{u_\varepsilon = 0\}$

by below at the origin we have to ensure that it stays below in a whole neighborhood. The size of this neighborhood is given by the first radius for which P separates at just one point at a distance $\delta_0\varepsilon$ from $x_n = 0$. Indeed, if r denotes the radius of this neighborhood then,

$$\|M\|r^2 \geq 2\delta_0\varepsilon \Rightarrow r^2 \geq \delta_0\|M\|^{-1}\varepsilon \geq \delta_0^2(\triangle P)^{-1}\varepsilon.$$

Hence, if $\varepsilon < \sigma_0(\delta_0)$ then P cannot touch by below $\{u_\varepsilon = 0\}$ at 0 in the r neighborhood.

Next we show that Proposition 7.5 follows from Proposition 7.4.

Assume by contradiction that for some

$$\varepsilon \leq \sigma_0(\delta_0) := \sigma^2(\frac{\delta_0^2}{4})$$

Γ touches $\{u_\varepsilon = 0\}$ by below at 0 in a $\delta_0(\triangle P)^{-\frac{1}{2}}\varepsilon^{\frac{1}{2}}$ neighborhood. By rescaling we find that

$$\left\{x_n = \frac{\varepsilon}{2}x'^T Mx'\right\}$$

touches by below $\{u = 0\}$ at 0 in a $\delta_0(\triangle P)^{-\frac{1}{2}}\varepsilon^{-\frac{1}{2}}$ neighborhood.

We apply Lemma 7.4 with

$$l = \frac{\delta_0}{2}(tr\ M)^{-\frac{1}{2}}\varepsilon^{-\frac{1}{2}}, \quad \theta = \frac{\delta_0^2}{4}, \quad M_1 = (tr\ M)^{-1}M, \quad \xi = 0,$$

and

$$\delta := \frac{\delta_0^2}{4}$$

thus,

$$\Gamma_1 = \left\{x_n = \frac{\varepsilon}{2}x'^T Mx'\right\}.$$

Since

$$\|M_1\| = (tr\ M)^{-1}\|M\| < \delta_0^{-1} < \delta^{-1},$$

$$\theta l^{-1} = \frac{\delta_0}{2}(tr\ M)^{\frac{1}{2}}\varepsilon^{\frac{1}{2}} < \varepsilon^{\frac{1}{2}} \leq \sigma(\frac{\delta_0^2}{4}) = \sigma(\delta)$$

we conclude

$$\delta \geq tr\ M_1 = 1$$

which is a contradiction.

Next we show that Theorem 7.3 implies Theorem 7.1. The proof is by compactness and is similar to the one from minimal surfaces. Assume by contradiction that there exist u_k, θ_k, l_k, ξ_k such that u_k is a local minimizer of J, $u_k(0) = 0$, the level set $\{u_k = 0\}$ stays in the flat cylinder

$$\{|x'| < l_k\} \times \{|x_n| < \theta_k\}$$

$\theta \geq \theta_0$, $\theta_k l_k^{-1} \to 0$ as $k \to \infty$ for which the conclusion of Theorem 7.1 doesn't hold.

Let A_k be the rescaling of the 0 level sets given by

$$(x', x_n) \in \{u_k = 0\} \mapsto (y', y_n) \in A_k$$

$$y' = x' l_k^{-1}, \quad y_n = x_n \theta_k^{-1}.$$

CLAIM 1. A_k has a subsequence that converges uniformly on $|y'| \leq 1/2$ to a set $A_\infty = \{(y', w(y')), \quad |y'| \leq 1/2\}$ where w is a Holder continuous function. In other words, given ε, all but a finite number of the A_k's from the subsequence are in an ε neighborhood of A_∞.

PROOF. Fix y_0', $|y_0'| \leq 1/2$ and suppose $(y_0', y_k) \in A_k$. We apply Theorem 7.3 for the function u_k in the cylinder

$$\{|x' - l_k y_0'| < l_k/2\} \times \{|x_n - \theta_k y_k| < 2\theta_k\}$$

in which the set $\{u_k = 0\}$ is trapped. Thus, there exist a universal constant $\eta_0 > 0$ and an increasing function $\varepsilon_0(\theta) > 0$, $\varepsilon_0(\theta) \to 0$ as $\theta \to 0$, such that $\{u_k = 0\}$ is trapped in the cylinder

$$\{|x' - l_k y_0| < \eta_0 l_k/2\} \times \{|x_n - \theta_k y_k| < 2(1 - \eta_0)\theta_k\}$$

provided that $4\theta_k l_k^{-1} \leq \varepsilon_0(2\theta_k)$. Rescaling back we find that

$$A_k \cap \{|y' - y_0'| \leq \eta_0/2\} \subset \{|y_n - y_k| \leq 2(1 - \eta_0)\}.$$

We apply the Harnack inequality repeatedly and we find that

$$A_k \cap \{|y' - y_0'| \leq \eta_0^m/2\} \subset \{|y_n - y_k| \leq 2(1 - \eta_0)^m\} \tag{7.2}$$

provided that

$$4\theta_k l_k^{-1} \leq \eta_0^{m-1} \varepsilon_0(2(1 - \eta_0)^m \theta_k).$$

Since these inequalities are satisfied for all k large we conclude that (7.2) holds for all but a finite number of k's.

There exist positive constants α, β depending only on η_0, such that if (7.2) holds for all $m \leq m_0$ then A_k is above the graph

$$y_n = y_k - 2(1 - \eta_0)^{m_0 - 1} - \alpha |y' - y_0'|^\beta$$

in the cylinder $|y'| \leq 1/2$.

Taking the supremum over these functions as y_0' varies we obtain that A_k is above the graph of a Hölder function $y_n = a_k(y')$. Similarly we obtain that A_k is below the graph of a Hölder function $y_n = b_k(y')$. Notice that

$$b_k - a_k \leq 4(1 - \eta_0)^{m_0 - 1} \tag{7.3}$$

and that a_k, b_k have a modulus of continuity bounded by the Hölder function αt^β. From Arzela-Ascoli Theorem we find that there exists a subsequence

a_{k_p} which converges uniformly to a function w. Using (7.3) we obtain that b_{k_p}, and therefore A_{k_p}, converge uniformly to w. $\square$

CLAIM 2. The function w is harmonic (in the viscosity sense).

PROOF. The proof is by contradiction. Fix a quadratic polynomial

$$y_n = P(y') = \frac{1}{2}y'^T My' + \xi \cdot y', \quad \|M\| < \delta^{-1}, \quad |\xi| < \delta^{-1}$$

such that $\triangle P > \delta$, $P(y') + \delta|y'|^2$ touches the graph of w, say, at 0 for simplicity, and stays below w in $|y'| < 2\delta$. Thus, for all k large we find points $(y_k{}', y_{kn})$ close to 0 such that $P(y') + const$ touches A_k by below at $(y_k{}', y_{kn})$ and stays below it in $|y' - y_k{}'| < \delta$.

This implies that, after eventually a translation, there exists a surface

$$\left\{x_n = \frac{\theta_k}{l_k^2}\frac{1}{2}x'^T Mx' + \frac{\theta_k}{l_k}\xi_k \cdot x'\right\}, \quad |\xi_k| < 2\delta^{-1}$$

that touches $\{u_k = 0\}$ at the origin and stays below it in the cylinder $|x'| < \delta l_k$. We write the above surface in the form

$$\left\{x_n = \frac{\delta^2\theta_k}{2(\delta l_k)^2}x'^T Mx' + \frac{\delta^2\theta_k}{\delta l_k}\delta^{-1}\xi_k \cdot x'\right\},$$

and we contradict Proposition 7.4 since $\theta_k \geq \theta_0$, $\theta_k l_k^{-1} \to 0$ and $\triangle P > \delta$.

Since w is harmonic, there exist $0 < \eta_1 < \eta_2$ small (depending only on n) such that

$$|w - \xi \cdot y'| < \eta_1/2 \quad \text{for } |y'| < 2\eta_2 \, .$$

Rescaling back and using the fact that A_k converge uniformly to the graph of w we conclude that for k large enough

$$\{u_k = 0\} \cap \{|x'| < 3l_k\eta_2/2\} \subset \{|x_n - \theta_k l_k^{-1}\xi \cdot x'| < 3\theta_k\eta_1/4\}.$$

This is a contradiction with the fact that u_k doesn't satisfy the conclusion of the Theorem 7.1. $\square$

8. A family of sliding surfaces

We prove Theorem 7.3 following the strategy of Section 5. The key step is to obtain estimates in measure. Since the level sets do not satisfy any equation on their own, one needs to look at their full collection.

In this section we introduce a family of sliding surfaces. These are rotation surfaces in $\mathbb{R}^{n+1}$ which we denote by $\mathcal{S}(Y, R)$. We say that the point Y is the center of $\mathcal{S}$ and R the radius. These surfaces are perturbations of the one dimensional solution. Roughly speaking they are obtained by first rotating $g_0(t)$ around the axis $t = -R$, and then modifying it outside the s level sets with $|s| < 1/2$, so that the resulting surface is a supersolution in the set $\{|x_{n+1}| \geq 1/2\}$.

The surfaces $\mathcal{S}$ in our setting play the same role as the paraboloids in Section 5. The key property that $\mathcal{S}$ satisfies is the following: Suppose that for fixed R, some surfaces $\mathcal{S}(Y,R)$ are tangent by above to the graph of u. Then the contact points project along e_n into a set with measure comparable with the measure of the projections of the centers Y along e_n (see Proposition 8.1).

We proceed with the explicit construction of $\mathcal{S}$.

For $|y_{n+1}| \le 1/4$, we define $\mathcal{S}(Y,R)$ as

$$\mathcal{S}(Y,R) := \{x_{n+1} = g_{y_{n+1},R}(H_0(y_{n+1}) + |x-y| - R)\}, \tag{8.1}$$

where the function $g_{s_0,R}$, respectively $h_{s_0,R}$, $H_{s_0,R}$ associated with it, are constructed below for $|s_0| \le 1/4$ and large R. For simplicity of notation we denote them by g, h, H.

Denote

$$\bar{C} = 1 + 8(n-1)\max\sqrt{W},$$

and let φ be such that

$$\frac{1}{\sqrt{2\varphi(s)}} = \frac{1}{\sqrt{2h_0(s)}} - \frac{\bar{C}_0}{R}(s-s_0),$$

where $\bar{C}_0$ is large enough so that the following holds

$$\begin{aligned} &\varphi(s) < h_0(s) - 2\bar{C}R^{-1}, \quad \text{if } s \in [-3/4,-1/2] \\ &\varphi(s) > h_0(s) + 2\bar{C}R^{-1}, \quad \text{if } s \in [1/2,3/4]. \end{aligned}$$

Let s_R near -1 be such that $h_0(s_R) = R^{-1}$, hence $1 + s_R \sim R^{-\frac{1}{2}}$. We define $h_{s_0,R} : [s_R, 1] \to \mathbb{R}$ as

$$h(s) = \begin{cases} h_0(s) - h_0(s_R) - \bar{C}R^{-1}(s-s_R) & \text{if } s \in [s_R, -\frac{1}{2}] \\ \varphi(s) \quad \text{if } s \in (-1/2,1/2) \\ h_0(s) + R^{-1} + \bar{C}R^{-1}(1-s) & \text{if } s \in [\frac{1}{2},1]. \end{cases}$$

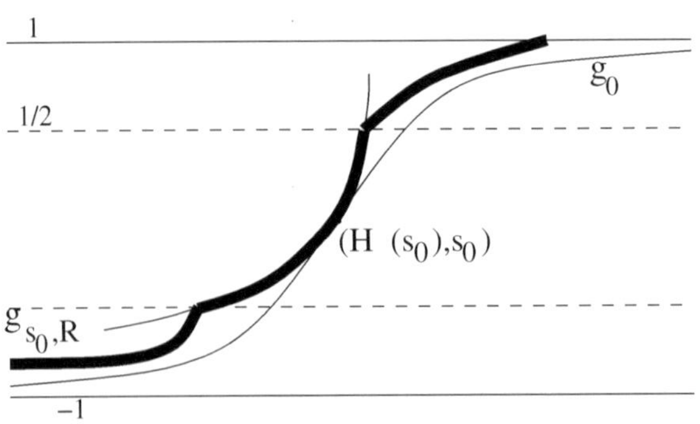

FIGURE 16. The profile of $\mathcal{S}(Y,R)$

For R large, $h(s) \geq c(1+s)(s-s_R)$ on $[s_R, 0]$, thus h is positive on $(s_R, 1]$. Define

$$H_{s_0,R}(s) = H_0(s_0) + \int_{s_0}^{s} \frac{1}{\sqrt{2h(\zeta)}} d\zeta$$

and for R large enough

$$H(s_R) \geq H_0(s_0) - \int_{s_R}^{s_0} \frac{1}{\sqrt{c(1+\zeta)(\zeta - s_R)}} d\zeta \geq -C \log R$$

$$H(1) \leq H_0(s_0) + \int_{s_0}^{1} \frac{1}{\sqrt{c(1-\zeta)^2 + R^{-1}}} d\zeta \leq C \log R.$$

Finally we define $g_{s_0,R}$ as

$$g_{s_0,R}(t) = \begin{array}{ll} s_R & \text{if } t < H(s_R) \\ H^{-1}(t) & \text{if } H(s_R) \leq t \leq H(1). \end{array}$$

Now we list some properties of the surfaces $\mathcal{S}(Y, R)$ (see Figure 16).

1) We have

$$h(s) > h_0(s) - 2\bar{C}R^{-1} > \varphi(s), \quad \text{if } s \in [-3/4, -1/2],$$
$$h(s) < h_0(s) + 2\bar{C}R^{-1} < \varphi(s), \quad \text{if } s \in [1/2, 3/4],$$

and

$$H(s) = H_0(s) - \frac{\bar{C}_0}{2R}(s - s_0)^2, \quad \text{if } |s| \leq 1/2,$$
$$H(s) > H_0(s) - \frac{\bar{C}_0}{2R}(s - s_0)^2 \quad \text{if } 1/2 < |s| < 3/4.$$

Let $\rho_{s_0,R}$ be the function whose graph is obtained from the graph of g_0 by the transformation

$$(t, s) \mapsto \left(t - \frac{\bar{C}_0}{2R}(s - s_0)^2, s \right) \quad \text{for } |s| < 3/4.$$

From the formulas above we obtain that $g = \rho$ for $|s| \leq 1/2$, and $g < \rho$ at all other points where ρ is defined. In other words, if $\mathbf{S}(Y, R)$ is the rotation surface

$$\mathbf{S}(Y, R) := \{x_{n+1} = \rho_{y_{n+1},R}(H_0(y_{n+1}) + |x - y| - R)\},$$

then, $\mathcal{S}(Y, R)$ coincides with $\mathbf{S}(Y, R)$ in the set $|x_{n+1}| \leq 1/2$ and stays below it at all the other points where $\mathbf{S}$ is defined.

Notice that

$$\mathbf{S}(Y, R) \subset \{|x_{n+1}| \leq 3/4\}$$

and it is defined only in a neighborhood of the sphere $|x - y| = R$ which is the y_{n+1} level set of $\mathbf{S}(Y, R)$.

2) We remark that $\mathcal{S}(Y,R)$ is constant s_R when

$$|x-y| \le R - C\log R,$$

and grows from s_R to 1 when

$$R - C\log R \le |x-y| \le R + C\log R.$$

3) If $s \in (s_R, -1/2) \cup (1/2, 1)$, then on the s level set of $\mathcal{S}$

$$\begin{aligned}\triangle \mathcal{S} \le & h'(s) + 2(n-1)R^{-1}\sqrt{2h(s)} \\ \le & h_0'(s) - \bar{C}R^{-1} + 4(n-1)R^{-1}\sqrt{h(s)} < h_0'(s),\end{aligned}$$

hence, $\mathcal{S}$ is a supersolution on its s level set. Moreover,

$$\lim_{s\to\pm 1/2^-} H'(s) < \lim_{s\to\pm 1/2^+} H'(s), \quad \lim_{s\to 1^-} H'(s) < \infty,$$

thus $\mathcal{S}(Y,R)$ is a strict supersolution for $|x_{n+1}| \ge 1/2$, that is $\mathcal{S}(Y,R)$ cannot touch by above the graph of a C^2 subsolution at a point X with $|x_{n+1}| \ge 1/2$.

4) If $|s| < 3/4$, then

$$|\varphi'(s) - h_0'(s)| \le CR^{-1}.$$

If v_S denotes the function with graph $\mathbf{S}(Y,R)$ defined above, one has

$$|\triangle v_S - h_0'(v_S)| \le C|\varphi' - h_0'| + CR^{-1} \le CR^{-1},$$

hence v_S is an approximate solution of the equation with a R^{-1} error.

Next proposition is quite surprising and is the key tool in proving Theorem 7.3. It is the analogue of the ABP estimate (Lemma 6.3) of Section 5.

Recall the notation $P_n := \{x_n = 0\}$.

PROPOSITION 8.1 (Measure estimate for contact points). *Let u be a C^2 subsolution i.e.*

$$\triangle u \ge h_0'(u), \quad |u| < 1.$$

Let A be a closed set in

$$P_n \cap \{|x_{n+1}| \le 1/4\}.$$

Assume that for each $Y \in A$ the surface $\mathcal{S}(Y + te_n, R)$, R large, stays above the graph of u when $t \to -\infty$ and, as t increases, it touches the graph by above for the first time at a point (contact point). If B denotes the projection of the contact points along e_n in P_n, then,

$$\bar{\mu}_0 |A| \le |B|$$

where $\bar{\mu}_0 > 0$ universal, small and $|A|$ represents the n-dimensional Lebesgue measure.

PROOF. Assume that $\mathcal{S}(Y, R)$ touches u by above at the point $X = (x, u(x))$. From the discussion above we find $|u(x)| < 1/2$.

Denote by ν the normal to the surface at X, i.e.

$$\nu = (\nu', \nu_{n+1}) = \frac{1}{\sqrt{1 + |\nabla u|^2}}(-\nabla u, 1).$$

For any contact point X the corresponding center Y is given by

$$Y(X) = (x + \frac{\nu'}{|\nu'|}\sigma, x_{n+1} + \omega) = F(X, \nu),$$

where

$$\omega = R\bar{C}_0^{-1}(\nu_{n+1}|\nu'|^{-1} - H_0'(x_{n+1}))$$
$$\sigma = -\frac{\bar{C}_0}{2R}\omega^2 + H_0(x_{n+1}) - H_0(x_{n+1} + \omega) + R.$$

The function F is smooth defined on

$$\{X \in \mathbb{R}^{n+1} : |x_{n+1}| < 1/2\} \times \{\nu \in \mathbb{R}^{n+1} : |\nu| = 1, c_1 < \nu_{n+1} < 1 - c_1\}.$$

The differential $D_X Y$ is a linear map defined on T_X, the tangent plane at X, and

$$D_X Y = F_X(X, \nu) + F_\nu(X, \nu) D_X \nu = F_X(X, \nu) - F_\nu(X, \nu) II_u$$

where II_u represents the second fundamental form of u at X. Writing the formula above for the surface $\mathbf{S}(Y, R)$ at X, we find

$$0 = F_X(X, \nu) - F_\nu(X, \nu) II_S$$

thus,

$$D_X Y = F_\nu(X, \nu)(II_S - II_u).$$

From (8.2) and (8.2), it is easy to check that

$$\|F_\nu(X, \nu)\| \le CR.$$

Since $\mathbf{S}$ touches u by above at X, we find that

$$D^2 v_S(x) - D^2 u(x) \ge 0,$$

where v_S is the function whose graph is $\mathbf{S}$. On the other hand, from (8.2),

$$\triangle v_S \le h_0'(x_{n+1}) + CR^{-1} = \triangle u + CR^{-1}$$

which implies

$$\|D^2 S(x) - D^2 u(x)\| \le CR^{-1}$$

or

$$\|II_S - II_u\| \le CR^{-1}.$$

This gives

$$\|D_X Y\| \le C.$$

The centers Z for which $X \in \mathbf{S}(Z, R)$ describe a rotation surface, around X. Note if $\mathbf{S}(\cdot, R)$ is above u, then its center is above this surface. The normal to the surface at $Y(X)$, which we denote by τ, belongs to the plane spanned by ν and e_{n+1}, and $c_2 < \tau_{n+1} < 1 - c_2$. Thus, we have

$$|\tau \cdot e_n| \le C|\nu \cdot e_n|$$

(notice that the tangent plane to the surface at $Y(X)$ is the range of $F_\nu(X, \nu)$.)

Let $\tilde{B}$ be the set of contact points, $\tilde{A}$ the set of the corresponding centers, $B = \pi_n \tilde{B}$ and $A = \pi_n \tilde{A}$. Remark that π_n is injective on $\tilde{A}$ and $\tilde{B}$ by construction. From above, we know that $\tilde{A}$ belongs to a Lipschitz surface. One has

$$\begin{aligned} |A| = \int_{\tilde{A}} |\tau(Y) \cdot e_n| dY &\le \int_{\tilde{B}} |\tau(Y) \cdot e_n||D_X Y| dX \\ &\le C \int_{\tilde{B}} |\nu(X) \cdot e_n| dX = C|B|, \end{aligned}$$

which is the desired claim. □

9. Proof of Theorem 7.3

In this section we prove Harnack inequality for minimizers of J. The main step is to show that if the 0 level set is close to the bottom $\{x_n = -\theta\}$ at one point, then the minimizer u is close to the one-dimensional solution $g_0(x_n + \theta)$ in a set of large measure. This set is obtained as the contact set between surfaces $\mathcal{S}(Y, R)$ and the graph of u. The measure estimate for the contact set comes from Proposition 8.1.

We estimate the size of the contact set by the Lebesgue measure of its projection along e_n. It is often convenient to think of the graph of u as a possibly multivalued graph over the set $P_n = \{x_n = 0\}$. Since the contact points occur only in the strip $\{|x_{n+1}| < 1/2\}$, we introduce the sets L and $Q_l \subset L$

$$\begin{aligned} L &= P_n \cap \{|x_{n+1}| \le 1/2\}, \\ Q_l &= L \cap \{|x'| \le l\}. \end{aligned}$$

Let us suppose we have a minimizer of J defined in a cylinder of size $2l$,

$$\{|x'| < 2l\} \times \{|x_n| < 2l\}$$

such that the 0 level set stays above $\{x_n = 0\}$ but close to it at one point above the origin

$$\begin{aligned} \{u = 0\} \cap \{|x'| < 2l\} &\subset \{x_n > 0\}, \\ (0, \theta) \in \{u = 0\}, &\quad \theta \ge \theta_0, \end{aligned}$$

for some fixed θ_0. We denote

$$\varepsilon := \frac{\theta}{l}.$$

Then the following result holds.

PROPOSITION 9.1. *There exist universal constants $\bar{C}_*$, $\bar{\mu}$, $\bar{c}_1$ such that if*

$$\bar{C}_*^k \varepsilon \le \bar{c}_1, \quad l \ge C(\theta_0)$$

then the set of points

$$(x, u(x)) \in \{|x'| \le l\} \times \{|x_{n+1}| \le 1/2\}$$

that satisfy

$$x_n \le \bar{C}_*^k \theta + H_0(u(x))$$

projects along e_n into a set of measure greater than $(1 - (1 - \bar{\mu})^k)|Q_l|$.

The inequality above says that the point $(x, u(x))$ stays at distance less than $\bar{C}_*^k \theta$ in the e_n direction from graph of the one dimensional solution $g_0(x_n)$.

This proposition is the analogue of Proposition 6.1 for minimal surfaces. The proof involves a number of technical lemmas that we describe below. We refer the reader to the original paper **[55]** for their proofs.

Since $(0, \theta)$ is close enough to the plane $\{x_n = 0\}$ we expect to find interior contact points with the sliding surfaces $\mathcal{S}(Y, R)$ for appropriate centers Y and radius R. This is explained in the next lemma. We remark that this step is straightforward in the case of minimal surfaces. The case of minimizers is more delicate since the bound on the 0 level does not guarantee that the contact points cannot occur on the boundary of the cylinder.

LEMMA 9.2 (Initial contact set). *If $\varepsilon \le \bar{c}_2$, then the points $(x, u(x))$ with the following properties*

1) $|x'| < l$, $|u(x)| < 1/2$
2) *there is a surface $\mathcal{S}(Y, R_0)$ that stays above u in the cylinder $|x'| < l$ and touches its graph at $(x, u(x))$, where*

$$R_0 = \frac{l^2}{\theta}, \quad l > \bar{C}_1 \log R_0$$

3) $\angle(\nu(x), \xi_0) \le l R_0^{-1}$
4) $x_n \le (1+b)\theta + H_0(u(x))$, $b < 1$ *is a positive fixed number project along e_n into a set of measure greater than $\bar{c}_2 (l\sqrt{b})^{n-1}$ provided that $l \ge C(b, \theta_0)$.*

Let $\tilde{D}_k$ represent the set of points on the graph of u that have by above a tangent surface $\mathcal{S}(Y, RC^{-k})$, where C is a large universal constant. Suppose that we have control on the e_n coordinate of these sets and denote by $D_k = \pi_n(\tilde{D}_k)$ their projections along e_n into L.

Recall that $\mathcal{S}(Y, RC^{-k})$ is an approximate solution of our equation with a $C^k R^{-1}$ error. If $\mathcal{S}(Y, RC^{-k})$ touches u by above at X_0 then, from the classical Harnack inequality for uniformly elliptic equations, the two surfaces stay $C^k R^{-1}$ close to each other in a neighborhood of X_0 (see Lemma 9.3). Thus, if we denote

$$E_k = \{Z \in L / \quad dist(Z, D_k) \leq C_1\},$$

then we control the e_n coordinate of a set on the graph of u that projects along e_n into E_k. Proposition 9.1 states that, in measure, E_k almost covers Q_l as k becomes larger and larger.

Looking at the graph of a solution between $-1/2$ and $1/2$ we see two different behaviors depending on the size of the neighborhood on which we focus our attention. Up to unit scale we see a Laplace equation and this is the reason why near a contact point the solution and the sliding surface stay very close to each other (see Lemma 9.3). At larger scales, the graph looks more like a minimal surface and the contact set behaves as in the minimal surface case. Lemma 9.4 below says that near (large scale) a point $Z \in D_k$ we can find a set of positive measure in D_{k+1}. Using a covering argument we show that the sets E_k "almost" cover Q_l as k increases.

Next we state the two lemmas that describe the different behaviors. We use the following notation:

$$\nu(x) := \frac{\nabla u}{|\nabla u|}(x) \in \mathbb{R}^n.$$

We assume that there exists a surface $\mathcal{S}(Y_0, R)$ that touches the graph of a solution u by above at a point $X_0 = (x_0, u(x_0))$ with

$$\angle(\nu(x_0), e_n) \leq \pi/4.$$

LEMMA 9.3 (Small scale measure estimate). *Given a constant $a > 1$ large, there exists $C(a) > 0$ depending also on a such that for each point $Z \in L \cap \mathcal{B}(\pi_n X_0, a)$ there exists x with*

1) $\pi_n(x, u(x)) = Z, \quad |x - x_0| \leq 2a,$
2) $(x - x_0) \cdot \nu(x_0) \leq H_0(u(x)) - H_0(u(x_0)) + C(a)R^{-1}.$

Next lemma corresponds to Lemma 6.3 in the minimal surface case.

LEMMA 9.4 (Large scale measure estimate). *Suppose that u is defined in the cylinder $\{|x'| < l\} \times \{|x_n| < l\}$ and satisfies the hypothesis above with*

$$|x_{0n}| < l/4, \quad |x_0'| = q, \quad q < l/4.$$

There exist constants $\bar{C}_1$, $\bar{C}_2$, such that if

$$q \geq \bar{C}_1, \quad R \geq l\bar{C}_1, \quad l \geq \bar{C}_1 \log R$$

then the set of points $(x, u(x))$ with the following four properties

1) $|x'| < q/15$, $|x - x_0| < 2q$, $|u(x)| < 1/2$,
2) *there is a surface $\mathcal{S}(Y, R/\bar{C}_2)$ that stays above u and touches its graph at $(x, u(x))$,*
3) $\angle(\nu(x), \nu(x_0)) \leq \bar{C}_1 q R^{-1}$,

4) $(x-x_0)\cdot\nu(x_0) \le \bar{C}_1 q^2 R^{-1} + H_0(u(x)) - H_0(u(x_0))$, *projects along* e_n *into a set of measure greater than* $q^{n-1}/\bar{C}_1$.

REMARK. The term $H_0(u(x)) - H_0(u(x_0))$ that appears in property 2 of Lemma 9.3 and property 4 of Lemma 9.4 represents the distance between the $u(x)$ level surface and the $u(x_0)$ level surface of a one dimensional solution.

Now we state the covering lemma that links the two scales.

LEMMA 9.5 (Covering lemma). *Let* D_k *be closed sets,* $D_k \subset L$, *with the following properties:*

1) $D_0 \cap Q_l \neq \emptyset$, $\quad D_0 \subset D_1 \subset D_2 \ldots$
2) *if* $Z_0 \in D_k \cap Q_{2l}$, $Z_1 \in L$, $|Z_1 - Z_0| = q$ *and* $2l \ge q \ge a$ *then,*

$$|D_{k+1} \cap \mathcal{B}(Z_1, q/10)| \ge \mu_1 |\mathcal{B}(Z_1, q) \cap L|$$

where $a > 1$ *(large),* μ_1 *(small) are given positive constants and* $l > 2a$.

Denote by E_k *the set*

$$E_k := \{Z \in L / \quad dist(Z, D_k) \le a\}.$$

Then there exists $\mu > 0$ *depending on* n, μ_1 *such that*

$$|Q_l \setminus E_k| \le (1-\mu)^k |Q_l|.$$

The proof of Proposition 9.1 follows from the lemmas above by showing that the sets D_k and E_k satisfy the hypotheses of the covering lemma above. The initial set D_0 is given by Lemma 9.2.

PROOF OF THEOREM 7.3. We assume that u is a local minimizer of J in the cylinder

$$\{|x'| < 2l\} \times \{|x_n| < 2l\},$$

and

$$u > 0 \quad \text{if } x_n > \theta, \qquad u < 0 \quad \text{if } x_n < -\theta, \qquad u(0) = 0.$$

We show that if the 0 level set is close to $x_n = -\theta$ at a point in $|x'| < l/4$ then the energy of u is large and obtain a contradiction.

First we apply Lemma 9.2 upside-down and obtain that there exists a small universal constant c_1 such that the points $(x, u(x))$ with

$$x_n \ge -\frac{\theta}{4} + H_0(u(x)), \quad |x'| \le \frac{l}{2}, \quad |u(x)| \le \frac{1}{2}$$

project along e_n in a set of measure greater than $c_1 l^{n-1}$, provided that $\theta l^{-1} \le \varepsilon_2(\theta_0)$ is small.

On the other hand, from Proposition 9.1 we find universal constants $\bar{C}_*$, $\bar{\mu}$ such that if

$$\{u=0\} \cap \{|x'|<l/4\} \cap \{x_n < (-1+\bar{C}_*^{-k_0}/4)\theta\} \neq \emptyset,$$
$$\theta l^{-1} \le \varepsilon_3(\theta_0, k_0)$$

then the set of points $(x, u(x))$ with

$$x_n \le -\frac{\theta}{2} + H_0(u(x)), \quad |x'| \le \frac{l}{2}, \quad |u(x)| \le \frac{1}{2}$$

project along e_n in a set of measure greater than $(1-(1-\bar{\mu})^{k_0})|Q_{l/2}|$. We show that if we choose k_0 large, universal such that

$$c_1 l^{n-1} > 2(1-\bar{\mu})^{k_0}|Q_{l/2}|$$

and ε small enough, then we obtain a contradiction.

As in Section 2 we use

$$\left(\frac{1}{2}|\nabla u|^2 + h_0(u)\right) dx' dx_n \ge \sqrt{2h_0(u)}|u_n| dx' dx_n$$
$$= \sqrt{2h_0(x_{n+1})} dx' dx_{n+1}.$$

Denote by

$$A_l := \{|x'|<l\} \times \{|x_n|<l\}.$$

Project along e_n the points $(x, u(x))$, with $|x_n| \le l$. From above we find that there exists a set of measure $c_1 l^{n-1}/2$ included in $Q_{l/2}$ in which these points project twice. Using barrier functions it is not difficult to show that the projection covers the band

$$\{|x_{n+1}| < 1 - c(l), \quad |x'|<l\}, \quad \text{and} \quad c(l) \to 0 \text{ as } l \to \infty.$$

Hence we can find two small universal constants c_2, $c_3 > 0$ such that

(9.1)
$$J(u, A_l) = \int_{A_l} \frac{1}{2}|\nabla u|^2 + h_0(u) dx$$
$$\ge \omega_{n-1} l^{n-1} \int_{c_2-1}^{1-c_2} \sqrt{2h_0(x_{n+1})} dx_{n+1} + \frac{c_1}{2}\left(\min_{|s|\le 1/2} \sqrt{2h_0(s)}\right) l^{n-1}$$
$$\ge \omega_{n-1} l^{n-1} \int_{-1}^{1} \sqrt{2h_0(s)} ds + c_3 l^{n-1},$$

where ω_{n-1} represents the volume of the $n-1$ dimensional unit sphere.

Assume by contradiction that there exist numbers l_k, θ_k with

$$\theta_k l_k^{-1} \to 0, \quad \theta_k \ge \theta_0,$$

and local minimizers u_k in A_{2l_k} satisfying the hypothesis of Theorem 7.3, and therefore also inequality (9.1).

Denote by $\varepsilon_k := l_k^{-1}$ and $v_k(x) := u_k(\varepsilon_k^{-1}x)$. From (9.1) we obtain

$$\begin{aligned} J_{\varepsilon_k}(v_k, A_1) &= \varepsilon_k^{n-1} J(u_k, A_{l_k}) \\ &\geq \omega_{n-1} \int_{-1}^{1} \sqrt{2h_0(s)}ds + c_3. \end{aligned}$$

On the other hand, as $k \to \infty$ we have

$$v_k \to \chi_E - \chi_{E^c} \quad \text{in } L^1_{loc}(A_1),$$

where $E = A_1 \cap \{x_n > 0\}$. By Theorem 2.1 one has

$$\lim_{k\to\infty} J_{\varepsilon_k}(v_k, A_1) = P_{A_1}E \int_{-1}^{1} \sqrt{2h_0(s)}ds = \omega_{n-1} \int_{-1}^{1} \sqrt{2h_0(s)}ds$$

and we reach a contradiction.

With this Theorem 7.3 is proved.

10. One-dimensional symmetry of global solutions

In this section we use the results of Section 7 and obtain 1D symmetry for phase transitions from the 1D symmetry of global minimal surfaces. Recall Corollary 7.2.

Corollary. *Let u be a global minimizer of J in $\mathbb{R}^n$. If the 0 level set $\{u = 0\}$ is asymptotically flat at ∞, then u is one-dimensional.*

At the end of Section 4 we used the minimal surface theory to show that the level sets of minimizers of J are asymptotically flat at ∞ in dimension $n \leq 7$. Using the corollary above we obtain

Theorem 10.1. *A global minimizer of*

$$J(u) = \int \frac{1}{2}|\nabla u|^2 + W(u)\,dx$$

is one-dimensional if $n \leq 7$.

Next we prove the following version of De Giorgi's conjecture.

Theorem 10.2. *Let $u \in C^2(\mathbb{R}^n)$ be a solution of*

$$\triangle u = W'(u), \tag{10.1}$$

such that

$$|u| \leq 1, \quad \partial_n u > 0, \quad \lim_{x_n \to \pm\infty} u(x', x_n) = \pm 1. \tag{10.2}$$

Then u is one-dimensional if $n \leq 8$.

Before we prove this Theorem we show that the ± 1 limit assumption implies that u is a global minimizer in $\mathbb{R}^n$.

Lemma 10.3. *If u satisfies (10.1), (10.2) then u is a global minimizer for J.*

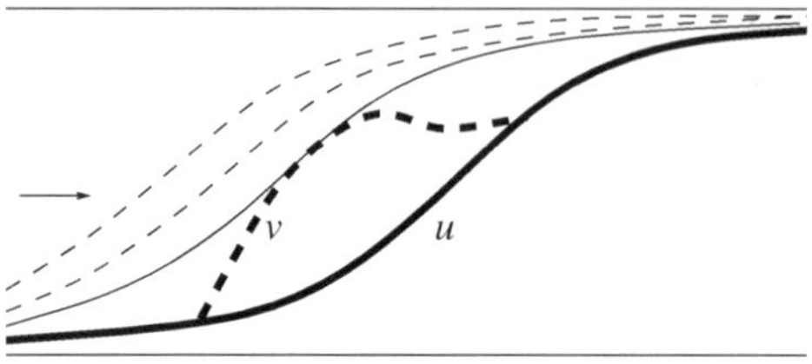

FIGURE 17. Sliding the graph of u

PROOF. It suffices to show that in B_R, u is the unique solution of

$$\triangle v = W'(v), \quad |v| < 1, \quad v = u \quad \text{on } \partial B_R.$$

Since

$$\lim_{x_n \to \infty} u(x', x_n) = 1,$$

we conclude that the graph of $u(x', x_n + t)$ is above the graph of v for large t. We slide this graph in the e_n direction till we touch v for the first time (see Figure 17). From the Strong Maximum Principle we find that the first touching point occurs on ∂B_R. Since u is strictly increasing in the e_n direction, we can slide the graph of $u(x', x_n + t)$ till it coincides with the graph of u, hence $u \geq v$. Similarly we obtain $u \leq v$ which proves that u is a global minimizer in $\mathbb{R}^n$. □

PROOF OF THEOREM 10.2. Assume $u(0) = 0$ and define $u_\varepsilon(x) = u(\varepsilon^{-1}x)$. Modica's theorem says that

$$u_{\varepsilon_k} \to \chi_E - \chi_{CE}$$

holds for some sequence $\varepsilon_k \to 0$. Moreover, $u_n > 0$ implies E^c is a subgraph, hence ∂E is a so-called quasi-solution in the e_n direction (see chapters 16, 17 in the book of Giusti [**40**]). Then ∂E is a hyperplane if $n \leq 8$ and the theorem follows from the corollary above.

□

If we assume that the 0 level set is a graph in the x_n direction which has a one sided linear bound at ∞ then the conjecture is true in any dimension.

THEOREM 10.4. *If u satisfies (10.1), (10.2) and*

$$\{u = 0\} \subset \{x_n < C(1 + |x'|)\}$$

then u is one-dimensional.

PROOF. As above, u is a global minimizer. Then the limiting minimal surface has a linear bound by above and again we can conclude (see Giusti [**40**]) that that ∂E is a hyperplane in any dimension. □

11. De Giorgi's conjecture in 2D and 3D

In this section we present proofs to De Giorgi's conjecture in 2D and 3D which were first obtained by Ghoussoub and Gui [**37**] and Ambrosio and Cabre [**5**] respectively.

Our approach is due to Farina [**30,31**] and it is based on a calculation of Sternberg and Zumbrun [**62,63**]. We follow it here since it closely resembles the one for minimal surfaces.

11.1. Minimal surfaces in 3D. Assume for simplicity that ∂E is a smooth minimal surface in $\mathbb{R}^n$. Denote by ν_x the normal to the surface at x and let ϕ be a function on ∂E with compact support. If we perturb ∂E by a map

$$x \mapsto x + \varepsilon\phi(x)\nu_x$$

then the area increases. By letting $\varepsilon \to 0$ we obtain (see Giusti [] for example) that the mean curvature is 0 at any point in ∂E and moreover

$$\int_{\partial E} \left(|\nabla\phi|^2 - \phi^2\kappa^2\right) d\mathcal{H}^{n-1} \geq 0,$$

where κ^2 represents the sum of the squares of the principal curvatures at $x \in \partial E$.

A smooth surface whose area increases under small local perturbations, i.e. has 0 mean curvature and satisfies the inequality above, is called a *stable minimal surface.*

One can obtain that a global minimal surface is a plane in $\mathbb{R}^3$ by choosing appropriate test functions ϕ. Indeed, if we choose ϕ_R with compact support in $\bar{B}_R$ as

$$\phi_R(x) := \begin{cases} 1, & |x| \leq 1 \\ \dfrac{1}{\log R}\log\dfrac{R}{|x|}, & 1 < |x| \leq R \end{cases}$$

then

$$\int_{\partial E} |\nabla\phi_R|^2 \, d\mathcal{H}^2 \leq (\log R)^{-2} \int_{\partial E\cap B_R} |x|^{-2} \, d\mathcal{H}^2.$$

From the minimality of ∂E we find that

$$\mathcal{H}^2(\partial E \cap B_r) \leq Cr^2$$

hence

$$\int_{\partial E\cap B_R} |x|^{-2} \, d\mathcal{H}^2 \leq C \log R.$$

In conclusion

$$\int_{\partial E} |\nabla\phi_R|^2 \, d\mathcal{H}^2 \to 0 \quad R \to \infty$$

which gives $\kappa = 0$ in $\partial E \cap B_1$. This implies $\kappa \equiv 0$, i.e. ∂E is a plane.

We remark that Simons' proof that global minimal cones are planar up to dimension 7, uses the stability condition by taking a test function $\phi(x)$ of the form $\kappa^2\psi(x)$.

11.2. Monotone solutions. First we check that monotone solutions of

$$\triangle u = W'(u),$$

are *stable*, i.e. the energy increases under small compact perturbations. Precisely, we show that

$$\int |\nabla\psi|^2 + W''(u)\psi^2\, dx \geq 0, \quad \forall \psi \in C_0^1(\mathbb{R}^n).$$

Indeed, assume u is increasing in the e_n direction, i.e. $u_n > 0$. Differentiating the equation we find

$$\triangle u_n = W''(u) u_n.$$

We multiply by ψ^2/u_n and integrate:

$$\int \nabla \frac{\psi^2}{u_n} \nabla u_n + W''(u)\psi^2\, dx = 0,$$

or

$$\int |\nabla\psi|^2 + W''(u)\psi^2\, dx = \int \left|\nabla\psi - \frac{\psi}{u_n}\nabla u_n\right|^2 dx \geq 0.$$

As in the minimal surface case, we would like to perturb the level sets in the normal direction infinitesimally by a function $\phi(x)$. This corresponds to taking $\psi := \phi|\nabla u|$ in the stability condition, i.e.

$$\int |\nabla(\phi|\nabla u|)|^2 + W''(u)\phi^2|\nabla u|^2\, dx \geq 0.$$

We obtain

$$\int |\nabla\phi|^2|\nabla u|^2 + \frac{1}{2}\nabla\phi^2\nabla(|\nabla u|^2) + \phi^2|\nabla(|\nabla u|)|^2 + \phi^2 W''(u)|\nabla u|^2\, dx \geq 0.$$

We integrate the second term by parts and use the equations for u_k,

$$\begin{aligned}\int \frac{1}{2}\nabla\phi^2\nabla(|\nabla u|^2)\, dx &= -\int \phi^2 \triangle(\frac{1}{2}|\nabla u|^2)\, dx \\ &= -\int \phi^2 \left(\|D^2 u\|^2 + W''(u)|\nabla u|^2\right) dx.\end{aligned}$$

By choosing a convenient system of coordinates it is not difficult to check that

$$\|D^2 u\|^2 - |\nabla|\nabla u||^2 = \kappa^2|\nabla u|^2 + |\nabla_T|\nabla u||^2$$

where ∇_T represents the tangential gradient along the level sets of u and $\kappa^2(x)$ represents the sum of the squares of the principal curvatures on the level set of u at x. In conclusion the stability inequality above simplifies as

$$\int \left(|\nabla \phi|^2 - \phi^2 \kappa^2\right) |\nabla u|^2 \, dx \geq 0.$$

This is precisely the stability condition for the minimal surface in which the $d\mathcal{H}^{n-1}$ measure on ∂E is replaced by the weight $|\nabla u|^2 dx$.

If the weight grows quadratically with the radius R in B_R i.e.

$$\int_{B_R} |\nabla u|^2 dx \leq CR^2,$$

then the same function ϕ_R gives as in the minimal surface case $\kappa \equiv 0$, that is u is one-dimensional.

This inequality is always true in $\mathbb{R}^2$ since L^∞ bounds for ∇u hold for bounded solutions under quite general assumptions on W, say for example W is Lipschitz. Therefore in $\mathbb{R}^2$ we obtain one-dimensionality from stability which, as we saw above, is a weaker condition than monotonicity.

THEOREM 11.1 (Ghoussoub-Gui). *Bounded stable global solution of*

$$\triangle u = W'(u), \quad \text{in } \mathbb{R}^2$$

are one-dimensional.

Ambrosio and Cabre proved that the quadratic bound holds also for monotone solutions in $\mathbb{R}^3$. As a consequence they obtain De Giorgi's conjecture in $\mathbb{R}^3$:

THEOREM 11.2 (Ambrosio-Cabre). *Bounded monotone global solution of*

$$\triangle u = W'(u), \quad \text{in } \mathbb{R}^3$$

are one-dimensional.

We sketch the proof. From (4.1) we see that the quadratic bound for the energy holds for minimizers in $\mathbb{R}^3$. Hence, if the limits are ± 1 as $x_3 \to \pm\infty$ then we are done by Lemma 10.3. Otherwise, denote by $\underline{u}(x')$, $\bar{u}(x')$ the limits of $u(x', x_3)$ as $x_3 \to \pm\infty$. It turns out that $\underline{u}$ and $\bar{u}$ are stable solutions of the same equation in $\mathbb{R}^2$, and by the theorem above they are one-dimensional. From this the bound for the energy follows.

We finally mention that the quadratic growth cannot hold in dimension $n > 3$ since it fails for the one-dimensional solution.

12. Fully nonlinear equations

The methods developed in Sections 7-10 are quite general and can be applied for other types of nonlinear, possibly degenerate elliptic equations.

With Valdinoci and Sciunzi we proved in [**66**] the same theorems for the energy

$$J_p(u,\Omega) := \int_\Omega \frac{1}{p}|\nabla u|^p + W(u)\,dx$$

and the corresponding p-Laplace equation

$$\triangle_p u = W'(u), \quad \triangle_p u := div(|\nabla u|^{p-2}|\nabla u|).$$

In this section we present $1D$ symmetry results for global solutions to fully nonlinear reaction-diffusion equations. We consider the equation $\mathbb{R}^n$

$$F(D^2u) = f(u), \tag{12.1}$$

where F is uniformly elliptic with ellipticity constants λ, Λ, $F(0) = 0$, and

$$f \in C^1([-1,1]), \quad f(\pm 1) = 0, \quad f'(-1) > 0, \quad f'(1) > 0.$$

The main assumption on F and f is that there exists a smooth increasing function $g_0 : \mathbb{R} \to [-1,1]$ (one dimensional solution) such that

$$\lim_{t \to \pm\infty} g_0(t) = \pm 1,$$

and g_0 solves the equation in all directions ξ, that is

$$F\left(D^2(g_0(x\cdot\xi))\right) = f\left(g_0(x\cdot\xi)\right)$$

for every unit vector $\xi \in \mathbb{R}^n$. In other words, the compatibility condition says that the values of F on the subset of rank one matrices $g_0''(t)\xi\otimes\xi$ equals $g(g_0(t))$.

For example if F is the maximal Pucci operator

$$F(M) = \mathcal{M}_{\lambda,\Lambda}(M) := \max_{\lambda I \le A \le \Lambda} tr(A\cdot M),$$

and $f(s) = \max\{\Lambda(s^3 - s), \lambda(s^3 - s)\}$ with

$$g_0(t) = \tanh\frac{t}{\sqrt{2}}.$$

One difficulty of the problem is that, unlike the semilinear case, there is no variational formulation of the problem. Another difficulty consists in the fact that it is not clear whether the blow-downs of $\{u = s\}$ satisfy any equation. We will see that in general the level sets satisfy at large scales a curvature equation depending on F, f.

Together with De Silva we proved in [**25**] that De Giorgi's conjecture holds in two dimensions for these nonlinear equations.

THEOREM 12.1 (De Silva-Savin). *Let $u \in C(\mathbb{R}^2)$ be a viscosity solution to (12.1) such that $|u| \le 1$ and $u_{x_2} > 0$. Then u is one-dimensional.*

The condition $u_{x_2} > 0$ can be relaxed to m monotonicity in the x_2 direction i.e.

$$u(x) < u(x + me_2), \quad \text{for some } m > 0.$$

The main step consists in proving that the sets $\{u < 0\}$ and $\{u > 0\}$ are convex. This is achieved by comparing u with appropriate explicit barriers. Then one can deduce that the 0 level set is included in a strip of arbitrarily small width.

The higher dimensional problem was analyzed in [**58**]. As in the semilinear case, we considered monotone viscosity solutions of (12.1) which "connect" the constant solutions -1 and 1 at $\pm\infty$,

$$u_{x_n} > 0, \qquad \lim_{x_n \to \pm\infty} u(x', x_n) = \pm 1, \tag{12.2}$$

Next result is a nonlinear version of the theorem of Modica. It says that if the blow-downs of the level sets converge to a limiting surface then this surface satisfies a curvature equation. This possible nonlinear curvature equation can be computed explicitly from F and f.

THEOREM 12.2 (Limiting equation). *Let u be a solution of (12.1), (12.2) and suppose that for a sequence $\varepsilon_k \to 0$, $\varepsilon_k\{u = 0\}$ converges uniformly on compact sets to a surface Σ. Then, there exists a function $\tilde{G}$ depending on F and g_0 such that*

$$\tilde{G}(\nu_\Sigma, II_\Sigma) = 0$$

in the viscosity sense where $\nu_\Sigma(x)$, $II_\Sigma(x)$ represent the upward normal and the second fundamental form of Σ at a point $x \in \Sigma$. The function $\tilde{G}(\xi, \cdot)$ is defined on the space of $n \times n$ symmetric matrices M with $M\xi = 0$, is homogenous of degree one, and uniformly elliptic.

The function $\tilde{G}$ is linear in the second argument and it depends on the derivatives of F on the n dimensional cone $\xi \otimes \xi$, $\xi \in \mathbb{R}^n$.

If F is invariant under rotations and $F \in C^1$ then $\tilde{G}(\xi, M) = trM$, i.e. Σ is a minimal graph as in the semilinear case. If for example $F = \mathcal{M}^+_{\lambda,\Lambda}$ then $\tilde{G}$ is nonlinear, $\tilde{G}(\xi, \cdot) = \mathcal{M}^-_{\lambda,\Lambda}$.

The methods developed in Sections 7-9 generalize to this setting. Due to the monotonicity assumption, the Harnack inequality requires only a one-sided bound.

THEOREM 12.3 (Harnack inequality). *Let u be a solution of (12.1), (12.2) with*

$$\{u = 0\} \cap \{|x'| < l\} \subset \{x \cdot \xi_0 > 0\},$$

for some unit vector

$$|\xi_0| = 1, \quad \angle(\xi_0, e_n) \le \beta < \pi/2.$$

Assume

$$(0, \theta) \in \{u = 0\}, \quad \theta \ge \theta_0$$

for some fixed θ_0. *If* $\theta/l \le \varepsilon(\theta_0)$ *then*

$$\{u = 0\} \cap \{|x'| < l/2\} \subset \{x \cdot \xi_0 < K\theta\},$$

where the constant K *depends only on* λ, Λ, n, f, g_0, β *and the constant* $\varepsilon(\theta_0) > 0$ *depends on the previous constants and* θ_0.

As a consequence of Harnack inequality we obtain a Liouville theorem for the level sets.

THEOREM 12.4. *If* $\{u = 0\}$ *is above (in the* e_n *direction) a plane* $\{x \cdot \xi = 0\}$, *then* u *is one-dimensional.*

A consequence of Theorem 12.4 is a proof of the Gibbons conjecture for equation (12.1). We recall that the conjecture states that global solutions are one-dimensional if the limits in (12.2) are uniform in x'.

If we assume more regularity on the operator F then we can use Theorem 12.3 and show the improvement of flatness for the level sets.

THEOREM 12.5 (Improvement of flatness). *Let* u *be a solution of (12.1), (12.2). Assume that* $F \in C^1$ *and*

$$0 \in \{u = 0\} \cap \{|x'| < l\} \subset \{|x \cdot \xi_0| < \theta\},$$

$$|\xi_0| = 1, \quad \angle(\xi_0, e_n) \le \beta < \pi/2, \quad \theta \ge \theta_0.$$

If $\theta/l \le \varepsilon(\theta_0)$ *then, for some unit vector* ξ_1

$$\{u = 0\} \cap \{|x'| < \eta_2 l\} \subset \{|x \cdot \xi_1| < \eta_1 \theta\},$$

where the constants $0 < \eta_1 < \eta_2 < 1$ *depend only on* λ, Λ, n *and the constant* $\varepsilon(\theta_0)$ *depends on* F, f, β *and* θ_0.

As a consequence of the theorem above we obtain $1D$ symmetry of global solutions which have one Lipschitz level set.

THEOREM 12.6. *Assume that* $F \in C^2$, u *solves (12.1), (12.2) and* $\{u = 0\}$ *is a Lipschitz graph in the* e_n *direction. Then* u *is one dimensional.*

References

[1] Alberti G., Ambrosio L., Cabre X., On a long-standing conjecture of E. De Giorgi: symmetry in 3D for general nonlinearities and a local minimality property. *Acta Appl. Math.* **65** (2001), no. 1-3, 9–33.

[2] Almgren F.J., The theory of varifolds- a variational calculus in the large for the k-dimensinal area integrand. Mimeographed notes, Princeton, 1965.

[3] Almgren F.J., Existence and regularity almost everywhere of solutions to elliptic variational problems among surfaces of varying topological type and singularity structure. *Ann. of Math.* (2) **87** (1968), 321–391.

[4] Alt H.W., Caffarelli L., Existence and regularity for a minimum problem with free boundary. *J. Reine Angew. Math* **325** (1981),105–144.

[5] Ambrosio L., Cabre X., Entire solutions of semilinear elliptic equations in $\mathbb{R}^3$ and a conjecture of De Giorgi. *J. American Math. Soc.* **13**, (2000), 725-739.

[6] Baldo S., Minimal interface criterion for phase transitions in mixtures of Cahn-Hilliard fluids. *Ann. Inst. H. Poincar Anal. Non Linaire*7 (1990), no. 2, 67–90.
[7] Barlow M., Bass R., Gui C., The Liouville property and a conjecture of De Giorgi. *Comm. Pure Appl. Math.* **53** (2000), no. 8, 1007–1038.
[8] Berestycki H., Caffarelli L., Nirenberg L., Further qualitative properties for elliptic equations in unbounded domains. *Ann. Scuola Norm. Sup. Pisa Cl. Sci. (4)* **25** (1997), no. 1-2, 69–94 (1998).
[9] Berestycki H., Hamel F., Monneau, R., One-dimensional symmetry of bounded entire solutions of some elliptic equations. *Duke Math. J.* **103** (2000), no. 3, 375–396.
[10] Bernstein S., Sur un theoreme de geometrie et son application aux equationes aux derivees partielles du type elliptique. *Comm. Soc. Math. de Kharkov* (2) **15** (1915-1917), 38–45. German transl.:*Math. Z.* **26** (1927), 551–558.
[11] Bombieri E., De Giorgi E., Giusti E., Minimal cones and the Bernstein problem. *Invent. Math.* **7** (1969), 243–268.
[12] Cahn J., Hilliard J., Free energy of a nonuniform system I. Interfacial free energy. *J. Chem. Phys* **28** (1958), 258–267.
[13] Cabre X., Terra J., Saddle-shaped solutions of bistable diffusion equations in all of $\mathbb{R}^{2m}$. Preprint, arXiv:0801.3379.
[14] Cabre X., Terra J., Qualitative properties of saddle-shaped solutions to bistable diffusion equations. Preprint, arXiv:0907.3008.
[15] Caffarelli L., Cabre X., Fully Nonlinear Elliptic Equations. Colloquium Publications 43, American Mathematical Society, Providence, RI, (1995).
[16] Caffarelli L., Cordoba A., An elementary regularity theory of minimal surfaces. *Differential Integral Equations* **6** (1993), no. 1, 1–13.
[17] Caffarelli L., Cordoba A., Uniform convergence of a singular perturbation problem. *Comm. Pure Appl. Math.* **48** (1995), no. 1, 1–12.
[18] Caffarelli L., Cordoba A., Phase transitions: uniform regulariy of the intermediate layers. *J. Reine Angew. Math.* **593** (2006), 209–235.
[19] Caffarelli L., Garofalo N., Segàla F., A gradient bound for entire solutions of quasilinear equations and its consequences. *Comm. Pure Appl. Math.* **47** (1994), no. 11, 1457–1473.
[20] De Giorgi E., Convergence problems for functional and operators. *Proc. Int. Meeting on Recent Methods in Nonlinear Analysis* (Rome, 1978), 131–188.
[21] De Giorgi E., Su una teoria generale della misura $(r-1)$-dimensionale in uno spazio ad r dimensioni. *Ann. Mat.Pura Appl.* (4) **36** (1954), 191–213.
[22] De Giorgi E., Nuovi teoremi relativi alle misure $(r-1)$-dimensionale inn uno spazio ad r dimensioni. *Ricerche Mat.* **4** (1955), 95–113.
[23] De Giorgi E., Complimenti alla teoria della misura $(n-1)$-dimensionale in uno spazio n-dimensionale. *Sem. Mat. Scuola Norm. Sup. Pisa*, 1960-61 . Editrice Tecnico Scientifica, Pisa, (1961).
[24] De Giorgi E., Frontiere orientate di misura minima. *Sem. Mat. Scuola Norm. Sup. Pisa 1960* Editrice Tecnico Scientifica, Pisa 1961.
[25] De Silva D., Savin O., Symmetry of global solutions to a class of fully nonlinear elliptic equations in 2D. *Indiana Univ. Math. J.,* **58** (1), (2009), 301–315.
[26] Del Pino M., Kowalczyk M., Wei J., On De Giorgi Conjecture in dimension $N \geq 9$. Preprint, arXiv:0806.3141.
[27] Douglas J., Solution of the problem of Plateau. *Trans. Amer. Math. Soc.* **33** (1931), 263–321.
[28] Farina A., Some remarks on a conjecture of De Giorgi. *Calc. Var. Partial Differential Equations* **8** (1999), 129–154.
[29] Farina A., Symmetry for solutions of semilinear elliptic equations in $\mathbb{R}^n$ and related conjectures. *Ricerche Mat.* **48** (1999), 129–154.

[30] Farina A., Propietes qualitatives de solutions d'equations et systemes d'equations non-lineaires. *Habilitation a diriger des recherches* 2002 Paris VI.

[31] Farina A., Sciunzi B., Valdinoci E., Bernstein and De Giorig type problems: new results via a geometric approach. Preprint (2007).

[32] Federer H., Some properties of distributions whose partial derivatives are represented by integration. *Bull. Amer. Math. Soc.* **74** (1968), 183–186.

[33] Federer H., Geometric measure theory. Springer-Verlag, Berlin-Heidelberg-New York, (1969).

[34] Federer H., The singular set of area minimizing rectifiable currents with codimension one and of area minimizing flat chains modulo two with arbitrary codimension. *Bull. Amer. Math. Soc.* **76** (1970), 767–771.

[35] Fleming W.H., On the oriented Plateau problem. *Rend. Circ. Mat. Palermo* (2) (1962) 69–90.

[36] Fonseca I., Tartar L., The gradient theory of phase transitions for systems with two potential wells. *Proc. Roy. Soc. Edinburgh Sect. A* **111** (1989), no. 1-2, 89–102.

[37] Ghoussoub N., Gui C., On a conjecture of De Giorgi and some related problems. *Math. Ann.* **311** (1998), 481-491.

[38] Ghoussoub N., Gui C., On De Giorgi's conjecture in dimensions 4 and 5. *Ann. of Math. (2)* **157** (2003), no. 1, 313–334.

[39] Gilbarg D., Trudinger N., Elliptic Partial Differential Equations of second order. Springer-Verlag, New York, (1983).

[40] Giusti E., Minimal Surfaces and functions of bounded variation. Birkhauser Verlag, Basel-Boston (1984).

[41] Jerison D., Monneau R., The existence of a symmetric global minimizer on $\mathbb{R}^{n-1}$ implies the existence of a counter-example to a conjecture of De Giorgi in $\mathbb{R}^n$. *C. R. Acad. Sci. Paris Sr. I Math* **.** **33**3 (2001), no. 5, 427–431.

[42] Kawohl B., Symmetry or not? *Math. Intelligencer* **20** (1998), 16–22.

[43] Kohn R., Sternberg P., Local minimisers and singular perturbations. *Proc. Roy. Soc. Edinburgh Sect. A* **111** (1989), no. 1-2, 69–84.

[44] Krylov N., Safonov M., A certain property of solutions of parabolic equations with measurable coefficients. *Izv. Akad. Nauk SSSR Ser. mat.,* **44** (1980), 161–175.

[45] Modica L., Γ-convergence to minimal surfaces problem and global solutions of $\Delta u = 2(u^3 - u)$. Proceedings of the International Meeting on Recent Methods in Nonlinear Analysis (Rome, 1978), pp. 223–244, Pitagora, Bologna, (1979).

[46] Modica L., A gradient bound and a Liouville theorem for nonlinear Poisson equations. *Comm. Pure Appl. Math.* **38** (1985), no. 5, 679–684.

[47] Modica L., The gradient theory of phase transitions and the minimal interface criterion. *Arch. Rational Mech. Anal.* **98** (1987), no. 2, 123–142.

[48] Modica L., Monotonicity of the energy for entire solutions of semilinear elliptic equations. Partial differential equations and the calculus of variations, Vol. II, 843–850, Progr. Nonlinear Differential Equations Appl., 2, Birkhuser Boston, Boston, MA, (1989).

[49] Modica L., Gradient theory of phase transitions with boundary contact energy. *Ann. Inst. H. Poincar Anal. No Lineaire* **4** (1987), no. 5, 487–512.

[50] Modica L., Mortola S., Some entire solutions in the plane of nonlinear Poisson equations. *Boll. Un. Mat. Ital. B (5)* **17** (1980), no. 2, 614–622.

[51] Modica L., Mortola S., Un esempio di Γ-convergenza. (Italian) *Boll. Un. Mat. Ital. B (5)* **14** (1977), no. 1, 285–299.

[52] Owen N., Nonconvex variational problems with general singular perturbations. *Trans. Amer. Math. Soc.* **310** (1988), no. 1, 393–404.

[53] Rado T., The problem of the least area and the problem of Plateau. *Math. Z.* **32** (1930), 763–796.

[54] Rado T., On the problem of Plateau, Springer-verlag, Berlin, (1933).

[55] Savin O., Phase transitions: Regularity of flat level sets. *PhD Thesis* UT Austin, (2003).
[56] Savin O., Regularity of flat level sets in phase transitions. *Ann. of Math.* (2) **169** (2009), no.1, 41–78.
[57] Savin O., Small perturbation solutions for elliptic equations. *Comm. Partial Differential Equations* **32** (2007), 557–578.
[58] Savin O., Entire solutions to a class of fully nonlinear elliptic equations. *Ann. Sc. Norm. Super. Pisa Cl. Sci.* (5), **7** (2008), no.3, 369–405.
[59] Simons J., Minimal varieties in riemannian manifolds. *Ann. of Math. (2)* **88** (1968), 62–105.
[60] Sternberg P., The effect of a singular perturbation on nonconvex variational problems. *Arch. Rational Mech. Anal.* **101** (1988), no. 3, 209–260.
[61] Sternberg P., Vector-valued local minimizers of nonconvex variational problems. *Rocky Mountain J. Math.* **21** (1991), no. 2, 799–807.
[62] Sternberg P., Zumbrun K., Connectivity of phase boundaries in a strictly convex domain. *Arch. rational Mech. Anal.* **141** (1998), 375–400.
[63] Sternberg P., Zumbrun K., A Poincare inequality with applications to volume constrained area minimizing surfaces. *J. Reine Angew. Math.* **503** (1998), 63–85.
[64] Tonegawa Y., Phase field model with a variable chemical potential. *Proc. Roy. Soc. Edinburgh Sect. A* **132** (2002), no. 4, 993–1019.
[65] Tonegawa Y., Domain dependent monotonicity formula for a singular perturbation problem. *Indiana Univ. Math. J.* **52** (2003), no. 1, 69–83.
[66] Valdinoci E., Sciunzi B., Savin O, Flat level set regularity of p-Laplace phase transitions. *Mem. Amer. Math. Soc.* **182** (2006), no. 858, vi+144pp.

Department of Mathematics, Columbia University, New York, NY 10027[1]
E-mail address: savin@math.columbia.edu

Current Developments in Mathematics, 2009

Recent Developments in Mathematical Quantum Chaos

Steve Zelditch

ABSTRACT. This is a survey of recent results on quantum ergodicity, specifically on the large energy limits of matrix elements relative to eigenfunctions of the Laplacian. It is mainly devoted to QUE (quantum unique ergodicity) results, i.e. results on the possible existence of a sparse subsequence of eigenfunctions with anomalous concentration. We cover the lower bounds on entropies of quantum limit measures due to Anantharaman, Nonnenmacher, and Rivière on compact Riemannian manifolds with Anosov flow. These lower bounds give new constraints on the possible quantum limits. We also cover the non-QUE result of Hassell in the case of the Bunimovich stadium. We include some discussion of Hecke eigenfunctions and recent results of Soundararajan completing Lindenstrauss' QUE result, in the context of matrix elements for Fourier integral operators. Finally, in answer to the potential question "why study matrix elements" it presents an application by the author to the geometry of nodal sets.

CONTENTS

1. Wave equation and geodesic flow 119
2. Weyl law and local Weyl law 132
3. Invariant states defined by eigenfunctions and their quantum limits 134
4. Quantum ergodicity and mixing of eigenfunctions 142
5. Concentration of eigenfunctions around hyperbolic closed geodesics 147
6. Boundary quantum ergodicity and quantum ergodic restriction 153
7. Hassell's scarring result for stadia 157
8. Matrix elements of Fourier integral operators 161
9. QUE of Hecke eigenfunctions 164

Research partially supported by NSF grant DMS 0904252.

10. Variance estimates: Rate of quantum ergodicity and mixing 169
11. Entropy of quantum limits on manifolds with Anosov geodesic flow 172
12. Applications to nodal hypersurfaces of eigenfunctions 194
References 200

Quantum chaos on Riemannian manifolds (M, g) is concerned with the asymptotics of eigenvalues and orthonormal bases of eigenfunctions

$$\Delta\varphi_j = \lambda_j^2\varphi_j, \quad \langle\varphi_j, \varphi_k\rangle = \delta_{jk} \tag{1}$$

of the Laplacian Δ when the geodesic flow $g^t : S_g^*M \to S_g^*M$ is ergodic or Anosov or in some other sense chaotic. Model examples include compact or finite volume hyperbolic surfaces $\mathbf{X}_\Gamma = \Gamma\backslash\mathbf{H}$, where $\mathbf{H}$ is the hyperbolic plane and $\Gamma \subset PSL(2, \mathbb{R})$ is a discrete subgroup. Of special interest are the arithmetic quotients when Γ is a discrete arithmetic subgroup. Other model example include Euclidean domains with ergodic billiards such as the Bunimovich stadium. The general question is, how does the dynamics of the geodesic (or billiard) flow make itself felt in the eigenvalues and eigenfunctions (or more general solutions of the wave equation), which by the correspondence principle of quantum mechanics must reproduce the classical limit as the eigenvalue tends to infinity?

To avoid confusion, we emphasize that we denote eigenvalues of the Laplacian by λ_j^2. They are usually viewed as energies $E_j = \lambda_j^2$. Their square roots λ_j are called the frequencies.

The most basic quantities testing the asymptotics of eigenfunctions are the matrix elements $\langle A\varphi_j, \varphi_j\rangle$ of pseudo-differential operators relative to the orthonormal basis of eigenfunctions. As explained below, these matrix elements measure the expected value of the observable A in the energy state φ_j. Much of the work in quantum ergodicity since the pioneering work of A. I. Schnirelman [**Sh.1**] is devoted to the study of the limits of $\langle A\varphi_j, \varphi_j\rangle$ as $\lambda_j \to \infty$. Although difficult to determine, these limits are still the most accessible aspects of eigenfunctions. A sequence of eigenfunctions is called ergodic or diffuse if the limit tends to $\int_{S_g^*M} \sigma_A d\mu_L$ where σ_A is the prinicipal symbol of A and $d\mu_L$ is the Liouville measure.

The first priority of this survey is to cover the important new results on the so-called scarring or QUE (quantum unique ergodicity) problem. Roughly speaking, the problem is whether every orthonormal basis becomes equidistributed in phase space with respect to Liouville measure as the eigenvalue tends to infinity, i.e.

$$\langle A\varphi_j, \varphi_j\rangle \to \int_{S_g^*M} \sigma_A d\mu_L$$

or whether there exists a sparse exceptional sequence which has a singular concentration. One series of positive results due to N. Anantharaman [**A**], N. Anantharaman-S. Nonnenmacher [**AN**] (see also [**ANK**]), and G. Rivière [**Riv**] flows from the study of entropies of quantum limit measures and are based on a difficult and complex microlocal PDE analysis of the long time behavior of the wave group on manifolds with Anosov geodesic flow. These results give lower bounds on entropies of quantum limit measures which imply that the limit measures must be to some extent diffuse. Another series of results due R. Soundararajan [**Sound1**] use L-function methods to complete the QUE result of E. Lindenstrauss [**LIND**] for Hecke eigenfunctions in the arithmetic case. In the negative direction, A. Hassell [**Has**] uses microlocal methods to proved the long-standing conjecture that generic stadia are not QUE but rather have exceptional sequences of bouncing ball type modes which concentrate or scar along a Lagrangian submanifold.

QUE is not the only interesting problem in quantum chaos. Another series of results in the arithmetic case are the remarkably sharp Luo-Sarnak asymptotics [**LS, LS2**] of the variances

$$\sum_{j:\lambda_j\leq\lambda} |\langle A\varphi_j, \varphi_j\rangle - \int_{S^*_g M} \sigma_A d\mu_L|^2$$

for holomorphic Hecke eigenforms and their recent generalization to smooth Maass forms by Zhao [**Zh**]. Physicists (see e.g. [**FP**]) have speculated that the variance is related to the classical auto-correlation function of the geodesic flow. The Luo-Sarnak-Zhao results partially confirm this conjecture in the special case of arithmetic surfaces and Hecke eigenfunctions, but correct it with an extra arithmetic factor.

A final direction we survey is the applications of quantum ergodicity to problems in nodal geometry. Given the large amount of work on matrix elements $\langle A\varphi_j, \varphi_j\rangle$, it is natural for geometric or PDE analysts to ask how one can use such matrix elements to study classical problems on eigenfunctions such as growth and distributions of zeros and critical points. One of our themes is that for real analytic (M, g), logarithms $\log|\varphi_j^{\mathbb{C}}(\zeta)|^2$ of analytic continuations of ergodic eigenfunctions are asymptotically "maximal plurisubharmonic functions" in Grauert tubes and as a result their zero sets have a special limit distribution [**Z5**]. Thus, ergodicity causes maximal oscillation in the real and complex domain and gives rise to special distributions of zeros and (conjecturally) to critical points.

In preparing this survey, we were only too aware of the large number of surveys and expository articles that already exist on the QUE and entropy problems, particularly [**AN2, ANK, CV3, LIND2, Sar3**] (and for earlier work [**Z1, Z2, M, Sar**]). To a large degree we closely follow the original sources, emphasizing intuition over the finer technical details. Although we go over much of the same material in what are now standard ways, we also go over some topics that do not seem as well known, and possibly could be improved, and we fill in some gaps in the literature. In particular, most expositions emphasize the quantum entropic uncertainty principle of [**AN**] due to its structural nature. So instead we emphasize the ideas of [**A**], which are less structural but in some ways more geometric.

One of the novelties is the study in §8 of matrix elements $\langle F\varphi_j, \varphi_j\rangle$ of Fourier integral operators F relative to the eigenfunctions. Hecke operators are examples of such F but restrictions to hypersurfaces provide a different kind of F which have recently come up in quantum ergodic restriction theory [**TZ3**]. We give a rather simple result on quantum limits for Fourier integral operators which answers the question, "what invariance property do quantum limit measures of Hecke eigenfunctions possess?". Hecke analysts have worked with a partial quasi-invariance principle due to Rudnick-Sarnak, and although the exact invariance principle may not simplify the known proofs of QUE it is of interest to know that one exists. We plan to put complete proofs in a future article.

Another aspect of eigenfunctions is their possible concentration of eigenfunctions around closed hyperbolic geodesic. Mass concentration within thin or shrinking tubes around such geodesics has been studied by S. Nonnenmacher-A.Voros [**NV**], Y. Colin de Verdière-B. Parisse [**CVP**], J. Toth and the author [**TZ2**] and outside of such tubes by N. Burq-M. Zworski [**BZ**] and H. Christianson [**Chr**]. Very striking and surprising studies of analogous behavior in the "quantum cat map" setting are in [**FNB, FN**].

A few words on the limitations of this survey. Due to the author's lack of knowledge, the L-function or arithmetic methods in QUE will not be discussed in more than a cursory way. Rather we concentrate on the PDE or microlocal aspects of quantum ergodicity. By necessity, microlocal analytical proofs must make direct connections between spectrum and geodesic flow, and cannot by-pass this obstacle by taking a special arithmetic route. Moreover, we restrict the setting to Laplacians on Riemannian manifolds for simplicity of exposition, but it is just a special case of the semi-classical asymptotics (as the Planck constant tends to zero) of spectra of Schrödinger operators $h^2\Delta+V$. Schrödinger eigenfunctions are equally relevant to physicists and mathematics. Another important setting is that of quantum "cat maps" on tori and other Kähler manifolds. In this context, the quantum dynamics is defined by quantizations of symplectic maps of Kähler manifolds acting on spaces of holomorphic sections of powers of a positive line bundle. The mathematics is quite similar to the Riemannian setting, and results in the cat map setting often suggest analogues in the Riemannian setting. In particular, the scarring results of Faure-Nonnemacher- de Bièvre [**FNB**] and the variance results of Kurlberg-Rudnick [**KR**] and Schubert [**Schu3**] are very relevant to the scarring and variance results we present in the Riemannian setting.

Although we do not discuss it here, a good portion of the literature of quantum chaos is devoted to numerics and physicists' heuristics. Some classics are [**Ber, He, He2**]. The article of A. Barnett [**Bar**] (and in the cat map case [**FNB, FN**]) is a relatively recent discussion of the numerical results in quantum ergodicity. There is a wealth of phenomenology standing behind the rigorous results and heuristic proofs in this field. Much of it still lies far beyond the scope of the current mathematical techniques.

Finally, we thank N. Anantharaman, A. Hassell, and S. Nonnenmacher for explanations of important points in their work. We have incorporated many of their clarifications in the survey. We also thank K. Burns, H. Hezari, J. Franks, E. Lindenstrauss, P. Sarnak, J. A. Toth, A. Wilkinson and J. Wunsch for further clarifications and corrections. Of course, responsibility for any errors that remain is the author's.

1. Wave equation and geodesic flow

Classical mechanics is the study of Hamiltonians (real valued functions) on phase space (a symplectic manifold) and their Hamiltonian flows. Quantum mechanics is the study of Hermitian operators on Hilbert spaces and the unitary groups they generate. The model quantum Hamiltonians we will discuss are Laplacians Δ on compact Riemannian manifolds (M, g) (with or without boundary). Throughout we denote the dimension of M by $d = \dim M$.

1.1. Geodesic flow and classical mechanics. The classical phase space in this setting is the cotangent bundle T^*M of M, equipped with its canonical symplectic form $\sum_i dx_i \wedge d\xi_i$. The metric defines the Hamiltonian $H(x, \xi) = |\xi|_g = \sqrt{\sum_{ij=1}^n g^{ij}(x)\xi_i\xi_j}$ on T^*M, where $g_{ij} = g(\frac{\partial}{\partial x_i}, \frac{\partial}{\partial x_j})$, $[g^{ij}]$ is the inverse matrix to $[g_{ij}]$. We denote the volume density of (M, g) by $dVol$ and the corresponding inner product on $L^2(M)$ by $\langle f, g\rangle$. The unit (co-) ball bundle is denoted $B^*M = \{(x, \xi) : |\xi| \leq 1\}$.

The classical evolution (dynamics) is given by the geodesic flow of (M, g), i.e. the Hamiltonian flow g^t of H on T^*M. Let X_H denote the Hamiltonian vector field of H. By definition, $g^t(x, \xi) = (x_t, \xi_t)$, where (x_t, ξ_t) is the terminal tangent vector at time t of the unit speed geodesic starting at x in the direction ξ. Here and below, we often identify T^*M with the tangent bundle TM using the metric to simplify the geometric description. The geodesic flow preserves the energy surfaces $\{H = E\}$ which are the co-sphere bundles $S^*_E M$. Due to the homogeneity of H, the flow on any energy surface $\{H = E\}$ is equivalent to that on the co-sphere bundle $S^*M = \{H = 1\}$.

We define the *Liouville measure* μ_L on S^*M to be the surface measure $d\mu_L = \frac{dx d\xi}{dH}$ induced by the Hamiltonian $H = |\xi|_g$ and by the symplectic volume measure $dx d\xi$ on T^*M. The geodesic flow (like any Hamiltonian flow) preserves $dx d\xi$. It also preserves Liouville measue on S^*M.

1.2. Ergodic and Anosov geodesic flows. The geodesic flow is called *ergodic* if the unitary operator

$$V^t a(\rho) = a(g^t(\rho)) \tag{2}$$

on $L^2(S^*M, d\mu_L)$ has no invariant L^2 functions besides constant functions. Equivalently, any invariant set $E \subset S^*M$ has either zero measure or full measure.

A geodesic flow g^t is called *Anosov* on on S_g^*M if the tangent bundle TS_g^*M splits into g^t invariant sub-bundles $E^u(\rho) \oplus E^s(\rho) \oplus \mathbb{R}\, X_H(\rho)$ where E^u is the unstable subspace and E^s the stable subspace. They are defined by

$$||dg^t v|| \leq Ce^{-\lambda t}||v||, \quad \forall v \in E^s, t \geq 0,$$

$$||dg^t v|| \leq Ce^{\lambda t}||v||, \quad \forall v \in E^u, t \leq 0.$$

The sub-bundles are integrable and give stable, resp. unstable foliations W^s, W^u. The leaves through x are denoted by $W^s(x), W^u(x)$. Thus, the geodesic flow contracts everything exponentially fast along the stable leaves and expands everything exponentially fast along the unstable leaves. Anosov geodesic flows are ergodic. We refer to [**Kl**] for background.

The unstable Jacobian $J^u(\rho)$ at ρ is defined by

$$J^u(\rho) = \det \left(dg^{-1}_{|E^u(g^1\rho)} \right). \tag{3}$$

Define

$$\begin{aligned} \Lambda &= - \sup_{\rho \in S^*M} \log ||J^u(\rho)|| \\ &= \inf_{\mu \in \mathcal{M}_I} \left| \int_{S^*M} \log J^u(\rho) d\mu(\rho) \right| > 0. \end{aligned} \tag{4}$$

If (M, g) is of dimension d and constant curvature -1 then $\Lambda = d - 1$. Here, $\mathcal{M}_I$ is the set of invariant probability measures for g^t.

The maximal expansion rate of the geodesic flow is defined by

$$\lambda_{\max} = \lim_{t \to \infty} \frac{1}{t} \log \sup_{\rho \in S^*M} ||dg^t_\rho||. \tag{5}$$

The derivative dg^t may be expressed in terms of Jacobi fields (see [**Kl**], Lemma 3.1.17). In the hyperbolic (Anosov case), the stable/unstable sub-bundles are spanned by unstable Jacobi fields $Y_1^u, \ldots, Y_{d-1}^u$ whose norms grow exponentially as $t \to +\infty$ and decay exponentially as $t \to -\infty$, resp. stable Jacobi fields $Y_1^s, \ldots, Y_{d-1}^s$ whose norms behave in the opposite fashion. The unstable Jacobian is the norm $||Y_1^u \wedge \cdots \wedge Y_{d-1}^u||$.

1.3. Eigenfunctions and eigenvalues of Δ. The quantization of the Hamiltonian H is the square root $\sqrt{\Delta}$ of the positive Laplacian,

$$\Delta = -\frac{1}{\sqrt{g}} \sum_{i,j=1}^{n} \frac{\partial}{\partial x_i} g^{ij} g \frac{\partial}{\partial x_j}$$

of (M, g). Here, $g = \det[g_{ij}]$. The eigenvalue problem on a compact Riemannian manifold

$$\Delta \varphi_j = \lambda_j^2 \varphi_j, \quad \langle \varphi_j, \varphi_k \rangle = \delta_{jk}$$

is dual under the Fourier transform to the wave equation. Here, $\{\varphi_j\}$ is a choice of orthonormal basis of eigenfunctions, which is not unique if the eigenvalues have multiplicities >1. The individual eigenfunctions are difficult to study directly, and so one generally forms the spectral projections kernel,

$$E(\lambda, x, y) = \sum_{j:\lambda_j \leq \lambda} \varphi_j(x)\varphi_j(y). \tag{6}$$

Semi-classical asymptotics is the study of the $\lambda \to \infty$ limit of the spectral data $\{\varphi_j, \lambda_j\}$ or of $E(\lambda, x, y)$.

1.4. Eigenvalues and Planck's constant. Although it is only a notational issue in this setting, it is conceptually useful to identify inverse frequencies with Planck's constant, i.e. to define

$$\hbar_j = \lambda_j^{-1}. \tag{7}$$

The eigenvalue problem then takes the form of a Schrödinger equation

$$\hbar_j^2 \Delta \varphi_j = \varphi_j.$$

Thus the high frequency limit is a special case of the semi-classical limit $\hbar \to 0$.

1.5. Eigenvalue Multiplicities. The multiplicity of an eigenvalue λ_j^2 is the dimension

$$m(\lambda_j) = \dim \ker(\Delta - \lambda_j^2) \tag{8}$$

of the eigenspace. As will be discussed in §2, the general Weyl law gives the bound $m(\lambda_j) \leq C\lambda_j^{d-1}$ in dimension d. In the case of negatively curved manifolds, the bound has been improved by a $(\log \lambda_j)^{-1}$ factor. Thus it is possible for sequences of eigenvalues to have the huge multiplicities $\lambda_j^{d-1}(\log \lambda_j)^{-1}$. When such multiplicities occur, it is possible to build up superpositions of eigenfunctions which have special concentration properties (see §5). Hassell's scarring theorem is based on the non-existence of such multiplicities for certain sequences of eigenvalues (§7).

In fact such multiplicities do occur in the quantum cap map setting for sparse sequences of eigenvalues and are responsible for the very strange behavior of eigenstates in that case [**FNB, FN**]. It is unknown if such sequences exist in the Riemannian Anosov setting. This is one of the fundamental obstacles to understanding whether or how scarring occurs in the Riemannian case.

1.6. Quantum evolution. Quantum evolution is given by the wave group

$$U^t = e^{it\sqrt{\Delta}},$$

which in a rigorous sense is the quantization of the geodesic flow Φ^t. It is generated by the pseudo-differential operator $\sqrt{\Delta}$, as defined by the spectral theorem (it has the same eigenfunctions as Δ and the eigenvalues λ_j). The (Schwartz) kernel of the wave group on a compact Riemannian manifold can be represented in terms of the spectral data by

$$U^t(x, y) = \sum_j e^{it\lambda_j} \varphi_j(x) \varphi_j(y), \tag{9}$$

or equivalently as the Fourier transform $\int_{\mathbb{R}} e^{it\lambda} dE(\lambda, x, y)$ of the spectral projections. Hence spectral asymptotics is often studied through the large time behavior of the wave group.

Evolution can be studied on the level of "points" or on the level of observables. Evolution of points in classical mechanics gives the orbits or trajectories of the geodesic flow (i.e. the parameterized geodesics). Evolution $\psi_t = U^t \psi$ of wave functions gives the "Schrödinger picture" of quantum mechanics. Eigenfunctions arise in quantum mechanics as stationary states, i.e. states ψ for which the probability measure $|\psi(t, x)|^2 dvol$ is constant where $\psi(t, x) = U^t \psi(x)$ is the evolving state. This follows from the fact that

$$U^t \varphi_k = e^{it\lambda_k} \varphi_k \tag{10}$$

and that $|e^{it\lambda_k}| = 1$. By unitarity $U^{t*} = U^{-t}$.

As an alternative one could define the quantum evolution as in [**AN**] to be the semi-classical Schrödinger propagator

$$U_\hbar^t := e^{-it\hbar\Delta}. \tag{11}$$

This replaces the homogeneous Fourier integral operator U^t by a semi-classical Fourier integral operator. The theories are quite parallel; we mainly keep to the homogeneous theory in this survey (See [**DSj**] and [**EZ**]) for background on these notions).

1.7. Observables. In the classical setting, observables are (real-valued) functions on T^*M which we usually take to be homogeneous of degree zero (or as zeroth order symbols). In the quantum setting, observables are ψDO's (pseudodifferential operators) of all orders; we often restrict to the subalgebra Ψ^0 of ψDO's of order zero. We denote by $\Psi^m(M)$ the subspace of pseudodifferential operators of order m. The algebra is defined by constructing a quantization Op from an algebra of symbols $a \in S^m(T^*M)$ of order m (polyhomogeneous functions on $T^*M\backslash 0$) to Ψ^m. A function on $T^*M\backslash 0$

is called polyhomogeneous or a classical symbol of order m, $a \in S^m_{phg}$, if it admits an asymptotic expansion,

$$a(x,\xi) \sim \sum_{j=0}^{\infty} a_{m-j}(x,\xi), \quad \text{where } a_k(x,\tau\xi) = \tau^k a_k(x,\xi), \ (|\xi| \geq 1)$$

for some m (called its order). The asympotics hold in the sense that

$$a - \sum_{j=0}^{R} a_{m-j} \in S^{m-R-1},$$

where $\sigma \in S^k$ if $\sup(1+|\xi|)^{j-k}|D_x^\alpha D_\xi^\beta \sigma(x,\xi)| < +\infty$ for all compact set and for all α, β, j. There is a semi-classical analogue in the $\hbar$ setting where the complete symbol $a_\hbar(x,\xi)$ depends on $\hbar$ in a similar poly-homogeneous way,

$$a_\hbar \sim \hbar^{-m} \sum_{j=0}^{R} \hbar^j a_{m-j}.$$

We refer to [**DSj, GSt, EZ, Dui**] for background on microlocal analysis (or semi-classical analysis), i.e. for the theory of pseudo-differential and Fourier integral operators.

The main idea is that such observables have good classical limits. The high frequency limit is mirrored in the behavior of the symbol as $|\xi| \to \infty$.

An useful type of observable is a smooth cutoff $\chi_U(x,\xi)$ to some open set $U \subset S^*M$ (it is homogeneous so it cuts off to the cone through U in T^*M). That is, a function which is one in a large ball in U and zero outside U. Then $Op(\chi_U)$ is called a microlocal cutoff to U. Its expectation values in the states φ_k give the phase space mass of the state in U or equivalently the probability amplitude that the particle is in U. (The modulus square of the amplitude is the probability).

There are several standard notations for quantizations of symbols a to operators A that we use in this survey. We use the notation $Op(a)$ or $\hat{a}$ interchangeably.

1.8. Heisenberg uncertainty principle. The Heisenberg uncertainty principle is the heuristic principle that one cannot measure things in regions of phase space where the product of the widths in configuration and momentum directions is $\leq \hbar$.

It is useful to microlocalize to sets which shrink as $\hbar \to 0$ using $\hbar$ (or λ)-dependent cutoffs such as $\chi_U(\lambda^\delta(x,\xi))$. The Heisenberg uncertainty principle is manifested in pseudo-differential calculus in the difficulty (or impossibility) of defining pseudo-differential cut-off operators $Op(\chi_U(\lambda^\delta(x,\xi))$ when $\delta > \frac{1}{2}$. That is, such small-scale cutoffs do not obey the usual rules of semi-classical analysis (behavior of symbols under composition). The uncertainty principle allows one to study eigenfunctions by microlocal methods in configuration space balls $B(x_0, \lambda_j^{-1})$ of radius $\hbar$ (since there is no constraint on the height) or in a phase space ball of radius $\lambda_j^{-1/2}$.

1.9. Matrix elements and Wigner distributions. These lectures are mainly concerned with the matrix elements $\langle Op(a)\varphi_j, \varphi_k\rangle$ of an observable relative to the eigenfunctions. The diagonal matrix elements

$$\rho_j(Op(a)) := \langle Op(a)\varphi_j, \varphi_j\rangle \tag{12}$$

are interpreted in quantum mechanics as the expected value of the observable $Op(a)$ in the energy state φ_j. The off-diagonal matrix elements are interpreted as transition amplitudes. Here, and below, an amplitude is a complex number whose modulus square is a probability.

If we fix the quantization $a \to Op(a)$, then the matrix elements can be represented by *Wigner distributions.* In the diagonal case, we define $W_k \in \mathcal{D}'(S^*M)$ by

$$\int_{S^*M} a dW_k := \langle Op(a)\varphi_k, \varphi_k\rangle. \tag{13}$$

1.10. Evolution of Observables. Evolution of observables is particularly relevant in quantum chaos and is known in physics as the "Heisenberg picture".

The evolution of observables in the Heisenberg picture is defined by

$$\alpha_t(A) := U^t A U^{-t}, \quad A \in \Psi^m(M). \tag{14}$$

and since Dirac's Principles of Quantum Mechanics, it was known to correspond to the classical evolution

$$V^t(a) := a \circ g^t \tag{15}$$

of observables $a \in C^\infty(S^*M)$. Egorov's theorem is the rigorous version of this correspondence: it says that α_t defines an order-preserving automorphism of $\Psi^*(M)$, i.e. $\alpha_t(A) \in \Psi^m(M)$ if $A \in \Psi^m(M)$, and that

$$\sigma_{U_t A U_t^*}(x, \xi) = \sigma_A(g^t(x, \xi)) := V^t(\sigma_A), \quad (x, \xi) \in T^*M \backslash 0. \tag{16}$$

This formula is almost universally taken to be the definition of quantization of a flow or map in the physics literature.

1.11. Hadamard parametrix. The link between spectral theory and geometry, and the source of Egorov's theorem for the wave group, is the construction of a parametrix (or WKB formula) for the wave kernel. For small times t, the simplest is the Hadamard parametrix,

$$U^t(x, y) \sim \int_0^\infty e^{i\theta(r^2(x,y) - t^2)} \sum_{k=0}^\infty U_k(x, y)\theta^{\frac{d-3}{2} - k} d\theta \quad (t < \mathrm{inj}(M, g)) \tag{17}$$

where $r(x, y)$ is the distance between points, $U_0(x, y) = \Theta^{-\frac{1}{2}}(x, y)$ is the volume 1/2-density, $inj(M, g)$ is the injectivity radius, and the higher Hadamard coefficients are obtained by solving transport equations along geodesics. The parametrix is asymptotic to the wave kernel in the sense of smoothness, i.e. the difference of the two sides of (17) is smooth. The relation (17) may be iterated using $(U^t)^m = U^{tm}$ to obtain a parametrix for long times. For manifolds without conjugate points, the Hadamard parametrix is globally well defined on the universal cover and can then be summed over the deck transformation group to obtain a parametrix on the quotient.

Egorov's theorem (16) may be proved by explicitly writing out the integral representation for $U^{-t}AU^t$ using (17) and by applying the stationary phase method to simplify the integral.

1.12. Ehrenfest time. The aim in quantum chaos is to obtain information about the high energy asymptotics as $\lambda_j \to \infty$ of eigenvalues and eigenfunctions by connecting information about U^t and g^t. The connection comes from comparing (9) and (17), or by using Egorov's theorem (16). But to use the hypothesis that g^t is ergodic or chaotic, one needs to exploit the connection as $\hbar \to 0$ and $t \to \infty$. The difficulty in quantum chaos is that the approximation of U^t by g^t is only a good one for t less than the *Eherenfest time*

$$T_E = \frac{\log |\hbar|}{\lambda_{\max}}, \tag{18}$$

where $\lambda_{\max}$ is defined in (5).

Roughly speaking, the idea is that the evolution of a well constructed "coherent" quantum state or particle is a moving lump that "tracks along" the trajectory of a classical particle up to time T_H and then slowly falls apart and stops acting like a classical particle. Numerical studies of long time dynamics of wave packets are given in works of E. J. Heller [**He, He2**] and rigorous treatments are in Bouzouina-Robert [**BouR**], Combescure-Robert [**CR**] and Schurbert [**Schu3**].

The basic result expressed in semi-classical notation is that there exists $\Gamma > 0$ such that

$$||U_\hbar^{-t} Op(a) U_\hbar^t - Op(a \circ g^t)|| \leq C\hbar\, e^{t\Gamma}. \tag{19}$$

Such an estimate has long been known if Γ is not specified in terms of the geodesics. It is implicit in [**Be**] and used explicitly in [**Z4**], based on long time wave equation estimates of Volovoy. The best constant should be $\Gamma = \lambda_{\max}$.

Thus, one only expects good joint asymptotics as $\hbar \to 0, T \to \infty$ for $t \leq T_E$. As a result, one can only exploit the approximation of U^t by g^t for the relatively short time T_E.

1.13. Modes, quasi-modes and coherent states. This survey is mainly concerned with eigenfunctions. But one of the main tools for studying eigenfunctions is the construction of quasi-modes or approximate eigenfunctions. Lagrangian and coherent states are closely related to quasi-modes. We refer to [**Dui, CV2, B.B, Ra, Ra2**] for background.

A quasi-mode $\{\psi_k\}$ of order zero is a sequence of L^2-normalized functions satisfying

$$||(\Delta - \mu_k)\psi_k||_{L^2} = O(1), \tag{20}$$

for a sequence of quasi-eigenvalues μ_k. By the spectral theorem, there must exist true eigenvalues in the interval $[\mu_k - \delta, \mu_k + \delta]$ for some $\delta > 0$. Moreover, if $E_{k,\delta}$ denotes the spectral projection for the Laplacian corresponding to this interval, then

$$||E_{k,\delta}\psi_k - \psi_k||_{L^2} = O(k^{-1}).$$

A more general definition is that a quasi-mode of order s is a sequence satisfying

$$||(\Delta - \mu_k)\psi_k||_{L^2} = O(\mu_k^{-s}). \tag{21}$$

Then

$$||E_{k,\delta}\psi_k - \psi_k||_{L^2} = O(k^{-s-1}).$$

This definition allows for very weak versions of quasi-modes, e.g. quasi-modes of order -1. They carry no information about eigenvalues since there always exist eigenvalues in intervals of width one. However, they do carry some information on eigenfunctions.

In references such as [**Dui**], quasi-modes are constructed in the form of oscillatory integrals or Lagrangian states,

$$\psi_\lambda(x) = \int_{\mathbb{R}^k} e^{i\lambda S(x,\xi)} a(x,\xi) d\xi.$$

The phase generates the Lagrangian submanifold

$$\Lambda_S = \{(x, d_x S) : d_\xi S(x,\xi) = 0\} \subset T^*M.$$

If $\Lambda \subset S^*M$ is a closed Lagrangian submanifold invariant under the geodesic flow then ψ_λ is a quasi-mode of order -1. In favorable circumstances, one can further determine the amplitude so that $||(-\Delta - \lambda^2)\psi_\lambda|| = O(\lambda^{-s})$ and construct a high order quasi-mode. Such a quasi-mode is the "quantization" of Λ and the associated sequence concentrates on Λ.

A coherent state is a special kind of Lagrangian state, or more precisely isotropic state, which quantizes a "point" in phase space rather than a Lagrangian submanifold. The simplest examples are the states $\varphi_\hbar^{x_0,\xi_0}(x) = \hbar^{-n/2} e^{\frac{i}{\hbar}\langle x,\xi_0\rangle} e^{-\frac{|x-x_0|^2}{\hbar}}$ on $\mathbb{R}^n$, which are localized to the smallest possible volume in phase space, namely a unit cube of volume $\hbar$ around (x_0, ξ_0). They are Lagrangian states with complex phases associated to positive Lagrangian submanifolds of complexified phase space.

1.14. Riemannian random waves. Eigenfunctions in the chaotic case are never similar to such quasi-modes or Lagrangian states. Rather, they are conjectured to resemble random waves. This heuristic was first proposed by M. V. Berry [**Ber**], who had in mind random Euclidean plane waves. The specific model discussed here was used by the author to show that eigenfunctions in the ergodic case behave to some degree like random orthonormal bases of spherical harmonics. We use the term Riemannian random waves for this model.

We decompose the spectrum of a compact Riemannian manifold (M, g) into intervals $I_N = [N, N+1]$ and define the finite dimensional Hilbert spaces $\mathcal{H}_{I_N}$ as the space of linear combinations of eigenfunctions with frequencies in an interval $I_N := [N, N+1]$. The precise decomposition of $\mathbb{R}$ into intervals is not canonical and any choice of intervals of finite width would work as long as the set of closed geodesics of (M, g) has measure zero. We will not discuss the other case since it has no bearing on quantum ergodicity. We denote by $\{\varphi_{Nj}\}_{j=1}^{d_N}$ an orthonormal basis of $\mathcal{H}_{I_N}$ where $d_N = \dim \mathcal{H}_{I_N}$. We endow the real vector space $\mathcal{H}_{I_N}$ with the Gaussian probability measure γ_N defined by

$$(22) \quad \gamma_N(s) = \left(\frac{d_N}{\pi}\right)^{d_N/2} e^{-d_N|c|^2} dc, \quad \psi = \sum_{j=1}^{d_N} c_j \varphi_{Nj}, \quad d_N = \dim \mathcal{H}_{I_N}.$$

Here, dc is d_N-dimensional real Lebesgue measure. The normalization is chosen so that $\mathbb{E}_{\gamma_N} \langle \psi, \psi \rangle = 1$, where $\mathbb{E}_{\gamma_N}$ is the expected value with respect to γ_N. Equivalently, the d_N real variables c_j $(j = 1, \dots, d_N)$ are independent identically distributed (i.i.d.) random variables with mean 0 and variance $\frac{1}{2d_N}$; i.e.

$$\mathbb{E}_{\gamma_N} c_j = 0, \quad \mathbb{E}_{\gamma_N} c_j c_k = \frac{1}{d_N} \delta_{jk}\,.$$

We note that the Gaussian ensemble is equivalent to picking $\psi_N \in \mathcal{H}_{I_N}$ at random from the unit sphere in $\mathcal{H}_{I_N}$ with respect to the L^2 inner product.

Numerical results confirming some features of the random wave model are given in [**HR**].

1.15. Hyperbolic plane. It is useful to have one concrete example in mind, and we will use the hyperbolic plane $\mathcal{H}$ or disc $\mathbf{D}$ as a running model. In the disc model $\mathbf{D} = \{z \in \mathbb{C}, |z| < 1\}$, the hyperbolic metric has the form $ds^2 = \frac{4|dz|^2}{(1-|z|^2)^2}$. Its isometry group is $G = PSU(1, 1)$; the stabilizer of 0 is $K \simeq SO(2)$ and thus $\mathbf{D} \simeq G/K$. Without comment we will also identify $\mathbf{D}$ with the upper half plane and G with $PSL(2, \mathbb{R})$. In hyperbolic polar coordinates centered at the origin 0, the (positive) Laplacian is the operator

$$-\Delta = \frac{\partial^2}{\partial r^2} + \coth r \frac{\partial}{\partial r} + \frac{1}{\sinh^2 r} \frac{\partial^2}{\partial \theta^2}.$$

The distance on $\mathbf{D}$ induced by the Riemannian metric will be denoted $d_{\mathbf{D}}$. We denote the volume form by $d(z)$.

The unit tangent bundle $S\mathbf{D}$ of the hyperbolic disc $\mathbf{D}$ may be identifed with the unit cosphere bundle $S^*\mathbf{D}$ by means of the metric. We further identify $S\mathbf{D} \equiv PSU(1,1)$, since $PSU(1,1)$ acts freely and transitively on $S\mathbf{D}$. We identify a unit tangent vector (z,v) with a group element g if $g \cdot (0,(1,0)) = (z,v)$.

We denote by $B = \{z \in \mathbb{C}, |z| = 1\}$ the boundary at infinity of $\mathbf{D}$. Then we can also identify $S\mathbf{D} \equiv \mathbf{D} \times B$. Here, we identify $(z,b) \in \mathbf{D} \times B$ with the unit tangent vector (z,v), where $v \in S_z\mathbf{D}$ is the vector tangent to the unique geodesic through z ending at b.

The geodesic flow g^t on $S\mathbf{D}$ is defined by $g^t(z,v) = (\gamma_v(t), \gamma_v'(t))$ where $\gamma_v(t)$ is the unit speed geodesic with initial value (z,v). The space of geodesics is the quotient of $S\mathbf{D}$ by the action of g^t. Each geodesic has a forward endpoint b and a backward endpoint b' in B, hence the space of geodesics of $\mathbf{D}$ may be identified with $B \times B \setminus \Delta$, where Δ denotes the diagonal in $B \times B$: To $(b', b) \in B \times B \setminus \Delta$ there corresponds a unique geodesic $\gamma_{b',b}$ whose forward endpoint at infinity equals b and whose backward endpoint equals b'. We then have the identification

$$S\mathbf{D} \equiv (B \times B \setminus \Delta) \times \mathbb{R}.$$

The choice of time parameter is defined – for instance – as follows: The point $(b', b, 0)$ is by definition the closest point to 0 on $\gamma_{b',b}$ and (b', b, t) denotes the point t units from $(b, b', 0)$ in signed distance towards b.

1.15.1. *Dynamics and group theory of* $G = SL(2,\mathbb{R})$. The generators of $sl(2,\mathbb{R})$ are denoted by

$$H = \begin{pmatrix} 1 & 0 \\ 0 & -1 \end{pmatrix}, \quad V = \begin{pmatrix} 0 & 1 \\ 1 & 0 \end{pmatrix}, \quad W = \begin{pmatrix} 0 & -1 \\ 1 & 0 \end{pmatrix}.$$

We denote the associated one parameter subgroups by A, A_-, K. We denote the raising/lowering operators for K-weights by

$$E^+ = H + iV, \quad E^- = H - iV. \tag{23}$$

The Casimir operator is then given by $4\,\Omega = H^2 + V^2 - W^2$; on K-invariant functions, the Casimir operator acts as the laplacian $\triangle$. We also put

$$X_+ = \begin{pmatrix} 0 & 1 \\ 0 & 0 \end{pmatrix}, \quad X_- = \begin{pmatrix} 0 & 0 \\ 1 & 0 \end{pmatrix},$$

and denote the associated subgroups by N, N_-. In the identification $S\mathbf{D} \equiv PSL(2,\mathbb{R})$ the geodesic flow is given by the right action $g^t(g) = ga_t$ of A, resp. the horocycle flow $(h^u)_{u \in \mathbb{R}}$ is defined by $h^u(g) = gn_u$, where

$$a_t = \begin{pmatrix} e^{t/2} & 0 \\ 0 & e^{-t/2} \end{pmatrix}, \quad n_u = \begin{pmatrix} 1 & u \\ 0 & 1 \end{pmatrix}.$$

The closed orbits of the geodesic flow g^t on $\Gamma\backslash G$ are denoted $\{\gamma\}$ and are in one-to-one correspondence with the conjugacy classes of hyperbolic elements of Γ. We denote by G_γ, respectively Γ_γ, the centralizer of γ in G, respectively Γ. The group Γ_γ is generated by an element γ_0 which is called a primitive hyperbolic geodesic. The length of γ is denoted $L_\gamma > 0$ and means that γ is conjugate, in G, to

$$a_\gamma = \begin{pmatrix} e^{L_\gamma/2} & 0 \\ 0 & e^{-L_\gamma/2} \end{pmatrix}. \tag{24}$$

If $\gamma = \gamma_0^k$ where γ_0 is primitive, then we call L_{γ_0} the primitive length of the closed geodesic γ.

1.15.2. *Non-Euclidean Fourier analysis.* Following [**Hel**], we denote by $\langle z, b\rangle$ the signed distance to 0 of the horocycle through the points $z \in \mathbf{D}$, $b \in B$. Equivalently,

$$e^{\langle z,b\rangle} = \frac{1-|z|^2}{|z-b|^2} = P_{\mathbf{D}}(z,b),$$

where $P_{\mathbf{D}}(z,b)$ is the Poisson kernel of the unit disc. We denote Lebesgue measure on B by $|db|$. We then introduce the hyperbolic plane waves $e^{(\frac{1}{2}+ir)\langle z,b\rangle}$, the analogues of the Euclidean plane waves $e^{i\langle x,\xi\rangle}$.

The non-Euclidean Fourier transform $\mathcal{F} : C_c\mathbf{D} \to C(B \times \mathbb{R}_+)$ is defined by

$$\mathcal{F}u(r,b) = \int_{\mathbf{D}} e^{(\frac{1}{2}-ir)\langle z,b\rangle} u(z) dVol(z).$$

The inverse Fourier transform is given by

$$\mathcal{F}^{-1}g(z) = \int_B \int_{\mathbb{R}} e^{(\frac{1}{2}+ir)\langle z,b\rangle} g(r,b) r \tanh(2\pi r) dr |db|.$$

The integration measure is the Plancherel measure. $\mathcal{F}$ extends to an isometry $\mathcal{F} : L^2(\mathbf{D}, dV) \to L^2(B \times \mathbb{R}_+, r\tanh(2\pi r) dr db)$.

1.15.3. *Wave kernel on* $\mathbf{D}$ *and wave kernel on* $\mathbf{X}_\Gamma$. The wave group $U^t = e^{it\sqrt{\Delta}}$ of the positive Laplacian on $\mathbf{D}$ is thus given in terms of Fourier analysis by (distribution) integral

$$U^t(z,w) = \int_B \int_{\mathbb{R}} e^{i(\frac{1}{4}+r^2)t} e^{(\frac{1}{2}+ir)\langle z,b\rangle} e^{(-\frac{1}{2}+ir)\langle w,b\rangle} r \tanh(2\pi r) dr |db|.$$

The integral over B may be eliminated by the stationary phase method to produce an integral over $\mathbb{R}$ alone; it is the Hadamard parametrix (17) in this case.

Now let $\Gamma \subset G$ be a discrete subgroup. In applications to quantum chaos we assume the quotient $\mathbf{X}_\Gamma$ is compact or of finite area.

The wave kernel on the quotient $\mathbf{X}_\Gamma$ is obtained by automorphizing the wave kernel on $\mathbf{D}$:

$$U_\Gamma^t(z,w) = \sum_{\gamma\in\Gamma} U_\Gamma^t(z,\gamma w). \tag{25}$$

This representation is one of the inputs into the Selberg trace formula. The elements of Γ are grouped into conjugacy classes $[\gamma] \in [\Gamma]$. The conjugacy classes correspond to closed geodesics of $\mathbf{X}_\Gamma$, i.e. periodic orbits of g^t on $S^*\mathbf{X}_\Gamma$.

Let

$$\Theta_\Gamma(T) = \#\{[\gamma] : L_\gamma \leq T\}. \tag{26}$$

The prime geodesic theorem asserts that

$$\Theta_\Gamma(T) \sim \frac{e^{h_{top}T}}{T}, \tag{27}$$

where h_{top} is the topological entropy of g^t. In fact, $h_{top} = 1$ for the hyperbolic case. The exponential growth of the length spectrum reflects the exponential growth of the geodesic flow.

The Ehrenfest time (18) is implicit in the exponential growth rate of $\Theta_\Gamma(T)$.

1.15.4. *Representation theory of G and spectral theory of $\triangle$.* Let $\Gamma \subset G$ be a co-compact discrete subgroup, and let us consider the automorphic eigenvalue problem on G/K:

$$\begin{cases} \triangle\varphi = (\frac{1}{4} + r^2)\varphi, \\ \varphi(\gamma z) = \varphi(z) \quad \text{for all } \gamma \in \Gamma \text{ and for all } z. \end{cases} \tag{28}$$

The solutions are the eigenfunctions of the Laplacian on the compact surface $\mathbf{X}_\Gamma = \Gamma\backslash\, G\, /K$. A standard notation for the eigenvalues is $\lambda^2 = s(1-s)$ where $s = \frac{1}{2} + ir$.

Eigenfunctions of the Laplacian are closely connected with the representation theory of G on G/Γ. We briefly describe the representation theory since the problems in quantum chaos have analogues in the discrete series as well as the unitary prinicipal series.

In the compact case, we have the decomposition into irreducibles,

$$L^2(\Gamma\backslash G) = \bigoplus_{j=1}^{S} \mathcal{C}_{ir_j} \oplus \bigoplus_{j=0}^{\infty} \mathcal{P}_{ir_j} \oplus \bigoplus_{m=2,\ m\ even}^{\infty} \mu_\Gamma(m)\mathcal{D}_m^+ \oplus \bigoplus_{m=2, m\ even}^{\infty} \mu_\Gamma(m)\mathcal{D}_m^-,$$

where $\mathcal{C}_{ir_j}$ denotes the complementary series representation, respectively $\mathcal{P}_{ir_j}$ denotes the unitary principal series representation, in which the Casimir operator $-\Omega = -(H^2 + V^2 - W^2)$ equals $s_j(1-s_j) = \frac{1}{4} + r_j^2$. In the complementary series case, $ir_j \in \mathbb{R}$ while in the principal series case $ir_j \in i\mathbb{R}^+$. The irreducibles are indexed by their K-invariant vectors $\{\varphi_{ir_j}\}$, which is assumed to be the given orthonormal basis of $\triangle$-eigenfunctions. Thus, the multiplicity of $\mathcal{P}_{ir_j}$ is the same as the multiplicity of the corresponding eigenvalue of $\triangle$.

Further, $\mathcal{D}_m^{\pm}$ denotes the holomorphic (respectively anti-holomorphic) discrete series representation with lowest (respectively highest) weight m, and $\mu_\Gamma(m)$ denotes its multiplicity; it depends only on the genus of $\mathbf{X}_\Gamma$. We denote by $\psi_{m,j}$ ($j = 1, \dots, \mu_\Gamma(m)$) a choice of orthonormal basis of the lowest weight vectors of $\mu_\Gamma(m)\mathcal{D}_m^+$ and write $\mu_\Gamma(m)\mathcal{D}_m^+ = \oplus_{j=1}^{\mu_\Gamma(m)} \mathcal{D}_{m,j}^+$ accordingly.

There is a a direct integral decomposition for co-finite subgroups such as $\Gamma = PSL(2, \mathbb{Z})$ or congruence subgroups. The non-compactness gives rise to a continuous spectral subspace of Eisenstein series and a discrete spectral subspace of cuspidal eigenfunctions (which is only known to be non-trivial in the case of arithmetic Γ).

1.15.5. *Helgason Poisson formula.* The Fourier transform of an L^2 function on $\mathbf{D}$ is an L^2 function on $B \times \mathbb{R}_+$. There is an extension of the Fourier transform and the inversion formula to tempered distributions. We only consider the case of Γ-automorphic eigenfunctions where Γ is co-compact. One then has $\mathcal{F}\varphi_{ir_j} = dT_{ir_j}(b) \otimes \delta_{r_j}(r)$, where dT_{ir_j} is a distribution on the ideal boundary B. The inversion formula is Helgason's Poisson formula,

$$\varphi_{ir}(z) = \int_B e^{(\frac{1}{2}+ir)\langle z,b\rangle} T_{ir,\varphi_{ir}}(db), \tag{29}$$

for all $z \in \mathbf{D}$. The kernel $e^{(\frac{1}{2}+ir)\langle z,b\rangle} = P_{\mathbf{D}}^{(\frac{1}{2}+ir)}(z,b)$ is called the generalized Poisson kernel. The distribution

$$T_{ir,\varphi_{ir}}(db) = \sum_{n\in\mathbb{Z}} a_n(r) b^n |db|. \tag{30}$$

is called the boundary value of φ_{ir} and is obtained from the Fourier expansion

$$\varphi_{ir}(z) = \sum_{n\in\mathbb{Z}} a_n(r)\Phi_{r,n}(z), \tag{31}$$

of φ_{ir} in the disc model in terms of the generalized spherical functions,

$$e^{(\frac{1}{2}+ir)\langle z,b\rangle} = \sum_{n\in\mathbb{Z}} \Phi_{r,n}(z) b^n, \quad b \in B. \tag{32}$$

Equivalently, the $\Phi_{r,n}$ are the joint eigenfunctions of Δ and of K.

When φ_{ir_j} is a Γ-invariant eigenfunction, the boundary values $T_{ir_j}(db)$ have the following invariance property:

$$\begin{aligned} \varphi_{ir_j}(\gamma z) = \varphi_{ir_j}(z) &\implies e^{(\frac{1}{2}+ir_j)\langle \gamma z,\gamma b\rangle} T_{ir_j}(d\gamma b) = e^{(\frac{1}{2}+ir_j)\langle z,b\rangle} T_{ir_j}(db) \\ &\implies T_{ir_j}(d\gamma b) = e^{-(\frac{1}{2}+ir_j)\langle \gamma\cdot 0,\gamma\cdot b\rangle} T_{ir_j}(db) \end{aligned} \tag{33}$$

This follows by the identities

$$\langle g \cdot z, g \cdot b \rangle = \langle z, b \rangle + \langle g \cdot 0, g \cdot b \rangle, \tag{34}$$

which implies

$$P_{\mathbf{D}}(gz, gb)\,|d(gb)| = P_{\mathbf{D}}(z, b)\,|db|. \tag{35}$$

An interesting heuristic related to Berry's random wave hypothesis (see §1.14) is the conjecture that the Fourier coefficients $a_n(r)$ in (30) should behave like independent Gaussian random variables of mean zero and variance one. This conjecture was stated explicitly and tested numerically by Aurich-Steiner in [**AS**]. Related tests of the random wave model for hyperbolic quotients are in Hejhal-Rackner [**HR**].

Otal [**O**] and Schmid [**Sch**] have shown that $T_{ir_j}(db)$ is the derivative of a Hölder $C^{\frac{1}{2}}$ continuous function F_{ir_j} on B. Since its zeroth Fourier coefficient is non-zero, $T_{ir_j}(db)$ is not literally the derivative of a periodic function, but it is the derivative of a function F_{ir_j} on $\mathbb{R}$ satisfying $F_{ir_j}(\theta+2\pi) = F_{ir_j}(\theta)+C_j$ for all $\theta \in \mathbb{R}$. Recall that for $0 \leq \delta \leq 1$ we say that a 2π-periodic function $F : \mathbb{R} \to \mathbb{C}$ is δ-Hölder if $|F(\theta) - F(\theta')| \leq C|\theta - \theta'|^{\delta}$. The smallest constant is denoted $||F||_{\delta}$ and Λ_{δ} denotes the Banach space of δ-Hölder functions, up to additive constants.

1.15.6. *Boundary values and representation theory* [**Z3**]. The distributions dT_{ir_j} have a natural interpretation in representation theory. We define $e_{ir_j} \in \mathcal{D}'(\Gamma\backslash PSL(2,\mathbb{R}))$ such that

$$e^{(\frac{1}{2}+ir_j)\langle z,b\rangle} T_{ir_j}(db) = e_{ir_j}(z,b) P(z,b) db. \tag{36}$$

The distribution e_{ir_j} is horocyclic-invariant and Γ-invariant. It may be expanded in a K-Fourier series,

$$e_{ir_j} = \sum_{n \in \mathbb{Z}} \varphi_{ir_j,n},$$

and it is easily seen (cf. [**Z2**]) that $\varphi_{ir_j,0} = \varphi_{ir_j}$ and that $\varphi_{ir_j,n}$ is obtained by applying the nth normalized raising or lowering operator (Maass operator) $E^{\pm} = H \pm iV$ to φ_{ir_j}. More precisely, one applies $(E^{\pm})^n$ and multiplies by the normalizing factor $\beta_{2ir_j,n} = \frac{1}{(2ir_j+1\pm 2n)\cdots(2ir_j+1\pm 2)}$.

2. Weyl law and local Weyl law

We now return to the general case. A basic result in semi-classical asymptotics is Weyl's law on counting eigenvalues. Intuitively, Weyl's law says that the trace of the spectral projector (6) has the asymptotics,

$$TrE_{\lambda} \sim \frac{Vol(|\xi|_g \leq \lambda)}{(2\pi)^d}, \tag{37}$$

where $d = \dim M$, where Vol is the symplectic volume measure relative to the natural symplectic form $\sum_{j=1}^{d} dx_j \wedge d\xi_j$ on T^*M. Thus, the dimension of the space where $H = \sqrt{\Delta}$ is $\leq \lambda$ is asymptotically the volume where its symbol $|\xi|_g \leq \lambda$. More precisely, on a manifold M (possibly with boundary),

$$N(\lambda) = \#\{j : \lambda_j \leq \lambda\}$$

$$= \frac{|B_d|}{(2\pi)^d} Vol(M, g)\lambda^d + \frac{|B_{d-1}|}{(2\pi)^{d-1}} Vol(\partial M, g)\lambda^{d-1} + R(\lambda), \tag{38}$$

$$\text{where} \quad R(\lambda) = O(\lambda^{d-1}).$$

Here, $|B_d|$ is the Euclidean volume of the unit ball and $Vol(M, g)$ (resp. $Vol(\partial M, g)$ is the volume of M (resp. ∂M) with respect to the metric g. The growth of the remainder term depends on the long time behavior of g^t. It is of order λ^{d-1} on the standard sphere or hemisphere, where all geodesics are periodic. By a classical theorem due to Duistermaat-Guillemin [**D.G**] (in the boundaryless case) and to Ivrii (in the boundary case),

$$R(\lambda) = o(\lambda^{d-1}), \quad \text{when the set of periodic geodesics has Liouville measure zero.}$$

The remainder is then of smaller order than the derivative of the principal term, and in the boundary case, of smaller order than the boundary contribution. In this generic case, one has asymptotics in shorter intervals:

$$N([\lambda, \lambda + 1]) = \#\{j : \lambda_j \in [\lambda, \lambda + 1]\} = n \frac{|B_d|}{(2\pi)^d} Vol(M, g)\lambda^{d-1} + o(\lambda^{d-1}). \tag{39}$$

Then the mean spacing between the eigenvalues in this interval is $\sim C_d Vol(M, g)^{-1}\lambda^{-(d-1)}$, where C_d is a constant depending on the dimension.

In the case of compact Riemannian manifolds of negative curvature without boundary, a sharper estimate of the remainder is possible:

$$R(\lambda) = O(\frac{\lambda^{d-1}}{\log \lambda}). \tag{40}$$

This remainder is proved using (25) and the exponential growth rate in (27). The estimate (40) was proved by Selberg in the case of compact hyperbolic quotients and was generalized to all compact Riemannian manifolds without conjugate points by Bérard [**Be**]. The logarithm in the remainder is a direct outcome of the fact that one can only use the geodesic approximation to U^t up to the Ehrenfest time T_E (18).

This estimate has not been improved in fifty years, and there are no better results in the constant curvature case than in the general negatively curved case. The remainder estimate does not rule out the implausible-seeming scenario that in the interval $[\lambda, \lambda + 1]$ there are only $\log \lambda$ distinct eigenvalues with multiplicities $\frac{\lambda^{d-1}}{\log \lambda}$. In fact, such implausibly large multiplicities do occur for a sparse set of "Planck constants" in the analogous case of quantizations of hyperbolic toral automorphisms (see [**FNB**]). The issue of possibly high multiplicity or clustering of eigenvalues is important in Hassell's scarring result as well as the Anantharaman-Nonnenmacher results on entropies.

An important generalization is the *local Weyl law* concerning the traces $TrAE(\lambda)$ where $A \in \Psi^m(M)$. When $\partial M = \emptyset$ (which we henceforth assume unless explicitly stating otherwise), it asserts that

$$\sum_{\lambda_j \leq \lambda} \langle A\varphi_j, \varphi_j \rangle = \frac{1}{(2\pi)^d} \left(\int_{B^*M} \sigma_A dx d\xi \right) \lambda^d + O(\lambda^{d-1}). \tag{41}$$

There is also a pointwise local Weyl law:

$$\sum_{\lambda_j \leq \lambda} |\varphi_j(x)|^2 = \frac{1}{(2\pi)^d} |B^d| \lambda^d + R(\lambda, x), \tag{42}$$

where $R(\lambda, x) = O(\lambda^{d-1})$ uniformly in x. When the periodic geodesics form a set of measure zero in S^*M, the remainders are of order $o(\lambda^{d-1})$, and one could average over the shorter interval $[\lambda, \lambda + 1]$. In the negatively curved case,

$$R(\lambda, x) = O(\frac{\lambda^{d-1}}{\log \lambda}).$$

Combining the Weyl and local Weyl law, we find the surface average of σ_A is a limit of traces:

$$\begin{aligned} \omega(A) &:= \frac{1}{\mu_L(S^*M)} \int_{S^*M} \sigma_A d\mu_L \\ &= \lim_{\lambda \to \infty} \frac{1}{N(\lambda)} \sum_{\lambda_j \leq \lambda} \langle A\varphi_j, \varphi_j \rangle \end{aligned} \tag{43}$$

Here, $d\mu_L$ is Liouville measure on S^*M (§1.1).

3. Invariant states defined by eigenfunctions and their quantum limits

When speaking of "states" in quantum mechanics, one might refer to a normalized wave function ψ or alternatively to the matrix elements $\rho_\psi(A) = \langle A\psi, \psi \rangle$ of an observable in the state. The latter use of "state" is a standard notion in C* algebras, and is central in quantum ergodicity and mixing. The states evidently have the properties: (i) $\rho_\psi(A^*A) \geq 0; (ii)\rho_\psi(I) = 1; (iii)\rho_\psi$ is continuous in the norm topology. The classical analogue is a probability measure, viewed as a positive normalized linear functional on $C(S^*M)$. In particular, eigenfunctions define states on Ψ^0 (more correctly, its closure in the norm topology) as in (12), which we repeat:

$$\rho_k(A) = \langle A\varphi_k, \varphi_k \rangle. \tag{44}$$

We could (and will) also consider "transition amplitudes" $\rho_{j,k}(A) = \langle A\varphi_j, \varphi_k \rangle$.

It is an immediate consequence of the fact that $U^t \varphi_j = e^{it\lambda_j}\varphi_j$ that the diagonal states ρ_k are invariant under the automorphism α_t (14):

$$\rho_k(U_t A U_t^*) = \rho_k(A). \tag{45}$$

In general, we denote by $\mathcal{E}$ the compact, convex set of states in the vector space of continuous linear functionals on the closure of Ψ^0 in its norm topology. We denote by

$$\mathcal{E}_{\mathbb{R}} = \{\rho \in \mathcal{E} : \rho \circ \alpha_t = \rho\} \tag{46}$$

the compact convex set of invariant states. For simplicity of notation, we continue to denote the closure by Ψ^0, and refer to [**Z6**] for a detailed exposition.

From a mathematical point of view, such the states $\rho_k(A)$ are important because they provide the simplest means of studying eigenfunctions: one tests eigenfunctions against observables by studying the values $\rho_k(A)$. Many standard inequalities in PDE (e.g. Carleman estimates) have the form of testing eigenfunctions against well chosen observables. Of course, the high eigenvalue asymptotics is not simple. One would like to know the behavior as $\lambda_j \to \infty$ (or $\hbar \to 0$ of the diagonal matrix elements $\langle A\varphi_j, \varphi_j \rangle$ and the transition amplitudes $\langle A\varphi_i, \varphi_j \rangle$ between states for $A \in \Psi^0(M)$. One of the principal problems of quantum chaos is the following:

PROBLEM 3.1. *Determine the set $\mathcal{Q}$ of "quantum limits", i.e. weak* limit points of the sequence of invariant eigenfunction states $\{\rho_k\}$ or equivalently of the Wigner distributions $\{W_k\}$.*

As will be illustrated below in simple examples, weak limits reflect the concentration and oscillation properties of eigenfunctions.

Off-diagonal matrix elements

$$\rho_{jk}(A) = \langle A\varphi_i, \varphi_j \rangle \tag{47}$$

are also important as transition amplitudes between states. They no longer define states since $\rho_{jk}(I) = 0$, are no longer positive, and are no longer invariant. Indeed, $\rho_{jk}(U_t A U_t^*) = e^{it(\lambda_j - \lambda_k)}\rho_{jk}(A)$, so they are eigenvectors of the automorphism α_t of (14). A sequence of such matrix elements cannot have a weak limit unless the spectral gap $\lambda_j - \lambda_k$ tends to a limit $\tau \in \mathbb{R}$. Problem 3.1 has the following extension to off-diagonal elements:

PROBLEM 3.2. *Determine the set $\mathcal{Q}_\tau$ of "quantum limits", i.e. weak* limit points of the sequence $\{W_{kj}\}$ of distributions on the classical phase space S^*M, defined by*

$$\int_X a dW_{kj} := \langle Op(a)\varphi_k, \varphi_j \rangle$$

where $\lambda_j - \lambda_k = \tau + o(1)$ *and where* $a \in C^\infty(S^*M)$, *or equivalently of the functionals* ρ_{jk}.

3.1. Simplest properties of quantum limits. The first is that $\langle K\varphi_k, \varphi_k\rangle \to 0$ for any compact operator K. Indeed, any orthonormal basis such as $\{\varphi_k\}$ tends to zero weakly in L^2. Hence $\{K\varphi_k\}$ tends to zero in L^2 for any compact operator and in particular the diagonal matrix elements tend to zero. It follows that for any $A \in \Psi^0(M)$, any limit of a sequence of $\langle A\varphi_k, \varphi_k\rangle$ is equally a limit of $\langle (A+K)\varphi_k, \varphi_k\rangle$.

Any two choices of Op, i.e. of quantizations of homogeneous symbols (of order zero) as pseudo-differential operators, are the same to leading order. Hence their difference is compact. Since a negative order pseudo-differential operator is compact, $\mathcal{Q}$ is independent of the definition of Op.

These properties do not use the fact that φ_j are eigenfunctions. The next property of the ρ_k are consequences of the fact $\rho_k \in \mathcal{E}_{\mathbb{R}}$.

PROPOSITION 3.3. $\mathcal{Q} \subset \mathcal{M}_I$, *where* $\mathcal{M}_I$ *is the convex set of invariant probability measures for the geodesic flow. They are also time-reversal invariant.*

To see this, we first observe that any weak * limit of of $\{\rho_k\}$ is a positive linear functional on symbols, which we identify with homogeneous functions of order zero on T^*M or with smooth functions on S^*M. Indeed, any limit of $\langle A\varphi_k, \varphi_k\rangle$, is bounded by $\inf_K ||A+K||$ (the infimum taken over compact operators), and for any $A \in \Psi^0$, $||\sigma_A||_{L^\infty} = \inf_K ||A+K||$. Hence any weak limit is bounded by a constant times $||\sigma_A||_{L^\infty}$ and is therefore continuous on $C(S^*M)$. It is a positive functional since each ρ_j is and hence any limit is a probability measure. The invariance under g^t follows from $\rho_k \in \mathcal{E}_{\mathbb{R}}$ by Egorov's theorem: any limit of $\rho_k(A)$ is a limit of $\rho_k(Op(\sigma_A \circ g^t))$ and hence the limit measure is g^t invariant. Furthermore, the limit measures are time-reversal invariant, i.e. invariant under $(x,\xi) \to (x,-\xi)$ since the eigenfunctions are real-valued.

Problem 3.1 is thus to identify which invariant measures in $\mathcal{M}_I$ show up as weak limits of the functionals ρ_k or equivalently of the Wigner distributions dW_k. The problem is that $\mathcal{M}_I$ can be a very large set. Examples of invariant probability measures for the geodesic flow include:

(1) Normalized Liouville measure $d\mu_L$. In fact, the functional ω of (43) is also a state on Ψ^0 for the reason explained above. A subsequence $\{\varphi_{j_k}\}$ of eigenfunctions is considered diffuse if $\rho_{j_k} \to \omega$.
(2) A periodic orbit measure μ_γ defined by $\mu_\gamma(A) = \frac{1}{L_\gamma}\int_\gamma \sigma_A ds$ where L_γ is the length of γ. A sequence of eigenfunctions for which $\rho_{k_j} \to \mu_\gamma$ obviously concentrates (or strongly "scars") on the closed geodesic.
(3) A finite convex combination $\sum_{j=1}^M c_j d\mu_{\gamma_j}$ of periodic orbit measures.
(4) A mixed measure such as $\frac{1}{2} d\mu_L + \frac{1}{2}\sum_{j=1}^M c_j d\mu_{\gamma_j}$

(5) A delta-function along an invariant Lagrangian manifold $\Lambda \subset S^*M$. The associated eigenfunctions are viewed as *localizing* along Λ.
(6) There are many additional kinds of singular measures in the Anosov case.

Thus, the constraint in the Proposition is far from enough to determine $\mathcal{Q}$.

To pin down $\mathcal{Q}$, it is necessary to find more constraints on the states ρ_k and their weak limits. We must use the quantum mechanics of ρ_k to pin down the possible classical limits.*

What possible additional constraints are there on the ρ_k? To date, only two are known.

- Further symmetries of the φ_j. In general there are none. But in special cases they exist. The most significant case is that of Hecke eigenfunctions, which carry an infinite number of further symmetries. See §9 and [**LIND, Sound1**].
- Entropies of limit measures. Lower bounds were obtained by N. Anantharaman and S. Nonnemacher [**A, AN**] in the case of (M, g) with Anosov geodesic flow (see also [**ANK, Riv**].)

These constraints are of a very different nature. The Hecke symmetries are special to arithmetic hyperbolic quotients, where one supplement the geodesic flow–wave group connection with the Hecke symmetries and connections to number theory. The entropy bounds are very general and only use the geodesic flow–wave group connection.

3.2. Ergodic sequences of eigenfunctions. A subsequence $\{\varphi_{j_k}\}$ of eigenfunctions is called *ergodic* if the only weak * limit of the sequence of ρ_{j_k} is $d\mu_L$ or equivalently the Liouville state ω. If $dW_{k_j} \to \omega$ then in particular, we have

$$\frac{1}{Vol(M)} \int_E |\varphi_{k_j}(x)|^2 dVol \to \frac{Vol(E)}{Vol(M)}$$

for any measurable set E whose boundary has measure zero. In the interpretation of $|\varphi_{k_j}(x)|^2 dVol$ as the probability density of finding a particle of energy λ_k^2 at x, this says that the sequence of probabilities tends to uniform measure.

However, $W_{k_j} \to \omega$ is much stronger since it says that the eigenfunctions become diffuse on the energy surface S^*M and not just on the configuration space M. One can quantize characteristic functions $\mathbf{1}_E$ of open sets in S^*M whose boundaries have measure zero. Then

$$\langle Op(\mathbf{1}_E)\varphi_j, \varphi_j\rangle = \text{the amplitude that the particle in energy state } \lambda_j^2 \text{ lies in E.}$$

For an ergodic sequence of eigenfunctions,

$$\langle Op(\mathbf{1}_E)\varphi_{j_k}, \varphi_{j_k}\rangle \to \frac{\mu_L(E)}{\mu_L(S^*M)},$$

so that the particle becomes diffuse, i.e. uniformly distributed on S^*M. This is the quantum analogue of the property of uniform distribution of typical geodesics of ergodic geodesic flows (Birkhoff's ergodic theorem).

3.3. QUE. The Laplacian Δ or (M, g) is said to be QUE (quantum uniquely ergodic) if $\mathcal{Q} = \{\mu_L\}$, i.e. the only quantum limit measure for any orthonormal basis of eigenfunctions is Liouville measure.

CONJECTURE 3.4. *(Rudnick-Sarnak,* [**RS**]*) Let (M, g) be a negatively curved manifold. Then Δ is QUE.*

In [**RS**], the case of arithmetic manifolds is investigated and the role of the Hecke operators is clarified and exploited. In particular the arithmetic QUE Conjecture refers to the limits of Hecke eigen-states, i.e. eigenfunctions of the arithmetic symmetries called Hecke operators. As reviewed in §9, in this case the conjecture in all its forms is more or less completely solved. E. Lindenstrauss [**LIND**] (together with the recent final step in the noncompact case by Soundararajan [**Sound1**]) proves this arithmetic QUE for arithmetic surfaces. If the multiplicity of the eigenvalues are uniformly bounded then one can deduce full QUE from arithmetic QUE, i.e. QUE for any orthonormal basis of eigenfunctions (see §9). Such uniform bounds on multiplicities are however far out of the range of current technology . The analogue of arithmetic QUE for Hecek holomorphic forms on noncompact arithmetic surfaces has recently been settled by Holowinsky and Soundararajan [**Hol, HS**]. In this case, the Hecke condition cannot be dropped due to the high multiplicity of such forms. Their methods are entirely arithmetical and we won't discuss them further here.

Although we are not discussing quantum cat maps in detail, it should be emphasized that quantizations of hyperbolic (Anosov) symplectic maps of the torus) are not QUE. For a sparse sequence of Planck constants $\hbar_k$, there exist eigenfunctions of the quantum cat map which partly scar on a hyperbolic fixed point (see Faure-Nonnenmacher-de Bièvre [**FNB**]). The multiplicities of the corresponding eigenvalues are of order $\hbar_k^{-1}/|\log \hbar|$. It is unknown if anything analogous can occur in the Riemannian setting, but as yet there is nothing to rule it out.

As pointed out in [**Z10**], QUE would follow if one could prove that quantum limits were invariant under a uniquely ergodic flow such as the horocycle flow of a compact hyperbolic quotient. A problem in trying to use this approach is that the horocycle flow is not Hamiltonian with respect to the standard symplectic form on $T^*\mathbf{X}_\Gamma$, i.e. it cannot be quantized. It is however Hamiltonian with respect to a modified symplectic structure. The modified symplectic structure corresponds to letting the weight of automorphic forms vary with the Δ eigenvalue. As a result, one does get large sequences of automorphic forms which are Liouville distributed. But QUE in the standard sense refers to sequences with fixed weight. It turns out that the weight has to grow so quickly for the QUE sequence that one cannot seem to relate the QUE result to quantum limits of forms with fixed weight [**Z10**].

3.4. Simplest example: S^1. The only computable example of ergodic eigenfunctions is the sequence of normalized eigenfunctions $2\sin k\pi x, 2\cos k\psi x$ on S^1. Note that they are real valued; the exponentials e^{ikx} are not quantum ergodic.

The energy surface of $T^*S^1 = S^1 \times \mathbb{R}_\xi$ is the pair of circles $|\xi| = \pm 1$. The geodesic flow has two invariant sets of positive measure (the two components), but the flow is time-reversal invariant under $(x, \xi) \to (x - \xi)$ and the quotient flow is ergodic. The real eigenfunctions are invariant under complex conjugation and therefore the quotient quantum system is ergodic.

Quantum ergodicity in this case amounts to

$$\frac{1}{\pi}\int_0^{2\pi} V(x)(\sin kx)^2 dx \to \frac{1}{2\pi}\int_0^{2\pi} V dx.$$

This is obvious by writing $\sin kx$ in terms of exponentials and using the Riemann-Lebesgue Lemma.

This simple example illustrates an important aspect of quantum limits: They are weak limits which owe to the fast oscillation of the eigenfunction squares. In the limit the oscillating functions tend to their mean values in the weak sense.

In particular, it illustrates why we only consider squares and not other powers of eigenfunctions: weak limits are not preserved under non-linear functionals such as powers. The study of L^p norms of chaotic eigenfunctions is very difficult (there are Iwaniec-Sarnak for Hecke eigenfunctions).

3.5. Example of quantum limits: the flat torus. The only examples where one can compute quantum limits directly are the completely integrable ones such as the standard sphere, torus or symmetric spaces. These examples of course lie at the opposite extreme from chaotic or ergodic dynamics. We use the simplest one, the flat torus $\mathbb{R}^n/\mathbb{Z}^n$, to illustrate the definition of weak* limits.

An orthonormal basis of eigenfunctions is furnished by the standard exponentials $e^{2\pi i\langle k,x\rangle}$ with $k \in \mathbb{Z}^n$. Obviously, $|e^{2\pi i\langle k,x\rangle}|^2 = 1$, so the eigenfunctions are diffuse in configuration space. But they are far from diffuse in phase space. For any pseudodifferential operator, $Ae^{2\pi i\langle k,x\rangle} = a(x,k)e^{2\pi i\langle k,x\rangle}$ where $a(x,k)$ is the complete symbol. Thus,

$$\langle Ae^{2\pi i\langle k,x\rangle}, e^{2\pi i\langle k,x\rangle}\rangle = \int_{\mathbb{R}^n/\mathbb{Z}^n} a(x,k)dx \sim \int_{\mathbb{R}^n/\mathbb{Z}^n} \sigma_A(x, \frac{k}{|k|})dx.$$

Thus, the Wigner distribution is $\delta_{\xi-k}$. A subsequence $e^{2\pi i\langle k_j,x\rangle}$ of eigenfunctions has a weak limit if and only if $\frac{k_j}{|k_j|}$ tends to a limit vector ξ_0 in the unit sphere in $\mathbb{R}^n$. In this case, the associated weak* limit is $\int_{\mathbb{R}^n/\mathbb{Z}^n}\sigma_A(x,\xi_0)dx$, i.e. the delta-function on the invariant torus $T_{\xi_0}\subset S^*M$ defined by the constant momentum condition $\xi=\xi_0$. The eigenfunctions are said to localize under this invariant torus for g^t.

The invariant torus is a Lagrangian submanifold of $T^*\mathbb{R}^n/\mathbb{Z}^n$, i.e. a submanifold of dimension n on which the standard symplectic form $dx\wedge d\xi$ restricts to zero. The exponentials are special cases of WKB or Lagrangian states $a_\hbar e^{\frac{i}{\hbar}S}$, where $a_\hbar$ is a semi-classical symbol $a_\hbar\sim a_j\hbar^j$. The associated Lagrangian submanifold is the graph $(x,dS(x))$ of S. Thus, $\langle k,x\rangle$ generates the Lagrangian submanifold $\xi=\frac{k}{|k|}$.

In general, one says that a sequence $\{\varphi_{j_k}\}$ of eigenfunctions concentrates microlocally on a Lagrangian submanifold $\Lambda\subset S^*M$ if

$$\langle A\varphi_{j_k},\varphi_{j_k}\rangle\to\int_\Lambda\sigma_A d\nu, \tag{48}$$

for some probability measure ν on Λ. Necessarily, Λ is invariant under g^t.

3.6. Hyperbolic case. Using the Helgason Poisson integral formula, the Wigner distributions can be expressed in terms of the dT_{ir_h} and e_{ir_j} as follows. As in [**Z3**], we define the hyperbolic calculus of pseudo-differential operators $Op(a)$ on $\mathbf{D}$ by

$$Op(a)e^{(\frac{1}{2}+ir)\langle z,b\rangle}=a(z,b,r)e^{(\frac{1}{2}+ir)\langle z,b\rangle}.$$

We assume that the complete symbol a is a polyhomogeneous function of r in the classical sense that

$$a(z,b,r)\sim\sum_{j=0}^{\infty}a_j(z,b)r^{-j+m}$$

for some m (called its order). By asymptotics is meant that

$$a(z,b,r)-\sum_{j=0}^{R}a_j(z,b)r^{-j+m}\in S^{m-R-1}$$

where $\sigma\in S^k$ if $\sup(1+r)^{j-k}|D_z^\alpha D_b^\beta D_r^j\sigma(z,b,r)|<+\infty$ for all compact set and for all α,β,j.

The non-Euclidean Fourier inversion formula then extends the definition of $Op(a)$ to $C_c^\infty(\mathbf{D})$:

$$Op(a)u(z)=\int_B\int_{\mathbb{R}}a(z,b,r)e^{(\frac{1}{2}+ir)\langle z,b\rangle}\mathcal{F}u(r,b)r\tanh(2\pi r)dr|db|.$$

A key property of Op is that $Op(a)$ commutes with the action of an element $\gamma\in G$ ($T_\gamma u(z)=u(\gamma z)$) if and only if $a(\gamma z,\gamma b,r)=a(z,b,r)$.

Γ-equivariant pseudodifferential operators then define operators on the quotient $\mathbf{X}_\Gamma$.

By the Helgason-Poisson formula one has a relative explicit formula for the Wigner distributions $W_{ir_j} \in \mathcal{D}'(S^*\mathbf{X}_\Gamma))$ defined by

(49)
$$\langle a, W_{ir_j}\rangle = \int_{S^*\mathbf{X}_\Gamma} a(g) W_{ir_j}(dg) := \langle Op(a)\varphi_{ir_j}, \varphi_{ir_j}\rangle_{L^2(\mathbf{X}_\Gamma)}, \quad a \in C^\infty(S^*\mathbf{X}_\Gamma).$$

Equivalently we have

(50)
$$W_{ir_j} = \varphi_{ir_j} e_{ir_j}.$$

Thus, the Wigner distributions are far more diffuse in phase space than in the case of the flat torus. But this does not rule out that their weak* limits could have a singular concentration.

3.6.1. *Patterson-Sullivan distributions.* Egorov's theorem implies that Wigner distributions tend to invariant measures for the geodesic flow. The question arises whether there exist g^t-invariant distributions constructed from eigenfunctions which are asymptotic to the Wigner distributions. In [**AZ**] such distributions were constructed on hyperbolic surfaces and termed "Patterson-Sullivan distributions" by analogy with their construction of boundary measures associated to ground states on infinite volume hyperbolic manifolds.

DEFINITION: The Patterson-Sullivan distribution associated to a real eigenfunction φ_{ir_j} is the distribution on $B \times B \setminus \Delta$ defined by

$$ps_{ir_j}(db', db) := \frac{T_{ir_j}(db) T_{ir_j}(db')}{|b - b'|^{1+2ir_j}}$$

If φ_{ir_j} is Γ-automorphic, then $ps_{ir_j}(db', db)$ is Γ-invariant and time reversal invariant.

To obtain a g^t-invariant distribution on $S^*\mathbf{X}_\Gamma$, we tensor ps_{ir_j} with dt. We then normalize by dividing by the integral against 1. The result is an invariant distribution $\hat{PS}_{ir_j}$ for g^t constructed as a quadratic expression in the eigenfunctions. In [**AZ**] Theorem 1.1, it is proved (theorem that

$$\int_{S^*\mathbf{X}_\Gamma} a\hat{PS}_{ir_j} = \int_{S^*\mathbf{X}_\Gamma} a W_{ir_j} + O(r_j^{-1}).$$

Hence the quantum limits problem is equally one of determining the weak* limits of the Patterson-Sullivan distributions. It is shown in [**AZ**] that they are residues of dynamical L-functions and hence have a purely classical definition.

In fact, there is an explicit intertwining operator L_{r_j} mapping $PS_{ir_j} \to W_{ir_j}$ and we have

(51)
$$\langle a, W_{ir_j}\rangle = \langle a, \hat{PS}_{ir_j}\rangle + r_j^{-1}\langle L_2(a), \hat{PS}_{ir_j}\rangle + O(r_j^{-2}).$$

4. Quantum ergodicity and mixing of eigenfunctions

In this section, we review a basic result on quantum ergodicity. We assume that the geodesic flow of (M, g) is ergodic. Ergodicity of g^t means that Liouville measure $d\mu_L$ is an ergodic measure for g^t on S^*M, i.e. an extreme point of $\mathcal{M}_I$. That is, there any g^t-invariant set has Liouville measure zero or one. Ergodicity is a spectral property of the operator V^t of (15) on $L^2(S^*M, d\mu_L)$, namely that V^t has 1 as an eigenvalue of multiplicity one. That is, the only invariant L^2 functions (with respect to Liouville measure) are the constant functions.

In this case, there is a general result which originated in the work of A. I. Schnirelman and was developed into the following theorem by S. Zelditch, Y. Colin de Verdière on manifolds without boundary and by P. Gérard-E. Leichtnam and S. Zelditch-M. Zworski on manifolds with boundary.

THEOREM 4.1. *Let (M, g) be a compact Riemannian manifold (possibly with boundary), and let $\{\lambda_j, \varphi_j\}$ be the spectral data of its Laplacian Δ. Then the geodesic flow G^t is ergodic on $(S^*M, d\mu_L)$ if and only if, for every $A \in \Psi^o(M)$, we have:*

- *(i)* $\lim_{\lambda\to\infty} \frac{1}{N(\lambda)} \sum_{\lambda_j \leq \lambda} |(A\varphi_j, \varphi_j) - \omega(A)|^2 = 0.$
- *(ii)* $(\forall\epsilon)(\exists\delta) \limsup_{\lambda\to\infty} \frac{1}{N(\lambda)} \sum_{\substack{j\neq k: \lambda_j, \lambda_k \leq \lambda \\ |\lambda_j - \lambda_k| < \delta}} |(A\varphi_j, \varphi_k)|^2 < \epsilon$

The diagonal result may be interpreted as a variance result for the local Weyl law. Since all the terms are positive, the asymptotic is equivalent to the existence of a s subsequence $\{\varphi_{j_k}\}$ of eigenfunctions whose indices j_k have counting density one for which $\langle A\varphi_{j_k}, \varphi_{j_k}\rangle \to \omega(A)$ for any $A \in \Psi^0(M)$. As above, such a sequence of eigenfunctions is called ergodic. One can sharpen the results by averaging over eigenvalues in the shorter interval $[\lambda, \lambda + 1]$ rather than in $[0, \lambda]$.

The off-diagonal statement was proved in [**Z9**] and the fact that its proof can be reversed to prove the converse direction was observed by Sunada in [**Su**]. A generalization to finite area hyperbolic surfaces is in [**Z8**].

The first statement (i) is essentially a convexity result. It remains true if one replaces the square by any convex function F on the spectrum of A,

$$\frac{1}{N(E)} \sum_{\lambda_j \leq E} F(\langle A\varphi_k, \varphi_k\rangle - \omega(A)) \to 0. \tag{52}$$

4.1. Quantum ergodicity in terms of operator time and space averages.

The diagonal variance asymptotics may be interpreted as a relation between operator time and space averages.

DEFINITION: *Let $A \in \Psi^0$ be an observable and define its time average to be:*

$$\langle A\rangle := \lim_{T\to\infty} \langle A\rangle_T,$$

where

$$\langle A \rangle_T := \frac{1}{2T} \int_{-T}^{T} U^t A U^{-t} dt$$

and its space average to be scalar operator

$$\omega(A) \cdot I$$

Then Theorem 4.1 (1) is (almost) equivalent to,

$$\langle A \rangle = \omega(A) I + K, \quad \text{where } \lim_{\lambda \to \infty} \omega_\lambda(K^* K) \to 0, \tag{53}$$

where $\omega_\lambda(A) = \frac{1}{N(\lambda)} Tr E(\lambda) A$. Thus, the time average equals the space average plus a term K which is semi-classically small in the sense that its Hilbert-Schmidt norm square $||E_\lambda K||_{HS}^2$ in the span of the eigenfunctions of eigenvalue $\leq \lambda$ is $o(N(\lambda))$.

This is not exactly equivalent to Theorem 4.1 (1) since it is independent of the choice of orthonormal basis, while the previous result depends on the choice of basis. However, when all eigenvalues have multiplicity one, then the two are equivalent. To see the equivalence, note that $\langle A \rangle$ commutes with $\sqrt{\Delta}$ and hence is diagonal in the basis $\{\varphi_j\}$ of joint eigenfunctions of $\langle A \rangle$ and of U_t. Hence K is the diagonal matrix with entries $\langle A\varphi_k, \varphi_k \rangle - \omega(A)$. The condition is therefore equivalent to

$$\lim_{E \to \infty} \frac{1}{N(E)} \sum_{\lambda_j \leq E} |\langle A\varphi_k, \varphi_k \rangle - \omega(A)|^2 = 0.$$

4.2. Heuristic proof of Theorem 4.1 (i). There is a simple picture of eigenfunction states which makes Theorem 4.1 seem obvious. Justifying the picture is more difficult than the formal proof below but the reader may find it illuminating and convincing.

First, one should re-formulate the ergodicity of g^t as a property of the Liouville measure $d\mu_L$: ergodicity is equivalent to the statement $d\mu_L$ is an extreme point of the compact convex set $\mathcal{M}_I$. Moreover, it implies that the Liouville state ω on $\Psi^0(M)$ is an extreme point of the compact convex set $\mathcal{E}_{\mathbb{R}}$ of invariant states for α_t of (14); see [**Ru**] for background. But the local Weyl law says that ω is also the limit of the convex combination $\frac{1}{N(E)} \sum_{\lambda_j \leq E} \rho_j$. An extreme point cannot be written as a convex combination of other states unless all the states in the combination are equal to it. In our case, ω is only a limit of convex combinations so it need not (and does not) equal each term. However, almost all terms in the sequence must tend to ω, and that is equivalent to (1).

One could make this argument rigorous by considering whether Liouville measure is an *exposed point* of $\mathcal{E}_I$ and $\mathcal{M}_I$. Namely, is there a linear functional Λ which is equal to zero at ω and is < 0 everywhere else on $\mathcal{E}_I$? If so, the fact that $\frac{1}{N(E)} \sum_{\lambda_j \leq E} \Lambda(\rho_j) \to 0$ implies that $\Lambda(\rho_j) \to 0$ for a subsequence of density one. For one gets an obvious contradiction if

$\Lambda(\rho_{j_k}) \leq -\epsilon < 0$ for some $\epsilon > 0$ and a subsequence of positive density. But then $\rho_{j_k} \to \omega$ since ω is the unique state with $\Lambda(\rho) = 0$.

In [**J**] it is proved that Liouville measure (or any ergodic measure) is exposed in $\mathcal{M}_I$. It is stated in the following form: For any ergodic invariant probability measure μ, there exists a continuous function f on S^*M so that μ is the unique f-maximizing measure in the sense that

$$\int f d\mu = \sup\left\{\int f dm : m \in \mathcal{M}_I\right\}.$$

To complete the proof, one would need to show that the extreme point ω is exposed in $\mathcal{E}_I$ for the C* algebra defined by the norm-closure of $\Psi^0(M)$.

4.3. Sketch of proof of Theorem 4.1 (i). We now sketch the proof of (52). By time averaging, we have

$$\sum_{\lambda_j \leq E} F(\langle A\varphi_k, \varphi_k\rangle - \omega(A)) = \sum_{\lambda_j \leq E} F(\langle\langle A\rangle_T - \omega(A)\varphi_k, \varphi_k\rangle). \tag{54}$$

We then apply the Peierls–Bogoliubov inequality

$$\sum_{j=1}^{n} F((B\varphi_j, \varphi_j)) \leq \operatorname{Tr} F(B)$$

with $B = \Pi_E[\langle A\rangle_T - \omega(A)]\Pi_E$ to get:

$$\sum_{\lambda_j \leq E} F(\langle\langle A\rangle_T - \omega(A)\varphi_k, \varphi_k\rangle) \leq \operatorname{Tr} F(\Pi_E[\langle A\rangle_T - \omega(A)]\Pi_E). \tag{55}$$

Here, Π_E is the spectral projection for $\hat{H}$ corresponding to the interval $[0, E]$. By the Berezin inequality (if $F(0) = 0$):

$$\begin{aligned} \tfrac{1}{N(E)} \operatorname{Tr} F(\Pi_E[\langle A\rangle_T - \omega(A)]\Pi_E) \; &\leq \; \tfrac{1}{N(E)} \operatorname{Tr} \Pi_E F([\langle A\rangle_T - \omega(A)])\Pi_E \\ &= \; \omega_E(\varphi(\langle A\rangle_T - \omega(A))). \end{aligned}$$

As long as F is smooth, $F(\langle A\rangle_T - \omega(A))$ is a pseudodifferential operator of order zero with principal symbol $F(\langle\sigma_A\rangle_T - \omega(A))$. By the assumption that $\omega_E \to \omega$ we get

$$\lim_{E\to\infty} \frac{1}{N(E)} \sum_{\lambda_j \leq E} F(\langle A\varphi_k, \varphi_k\rangle - \omega(A)) \leq \int_{\{H=1\}} F(\langle\sigma_A\rangle_T - \omega(A)) d\mu_L.$$

As $T \to \infty$ the right side approaches $\varphi(0)$ by the dominated convergence theorem and by Birkhoff's ergodic theorem. Since the left hand side is independent of T, this implies that

$$\lim_{E\to\infty} \frac{1}{N(E)} \sum_{\lambda_j \leq E} F(\langle A\varphi_k, \varphi_k\rangle - \omega(A)) = 0$$

for any smooth convex F on $\operatorname{Spec}(A)$ with $F(0) = 0$. $\square$

This proof can only be used directly for scalar Laplacians on manifolds without boundary, but it still works as a template in more involved situations. For instance, on manifolds with boundary, conjugation by the wave group is not a true automorphism of the observable algebra. In quantum ergodic restriction theorems (see §6), the appropriate conjugation is an endomorphism but not an automorphism. Or when Δ has continuous spectrum (as in finite area hyperbolic surfaces), one must adapt the proof to states which are not L^2-normalized [**Z8**].

4.4. QUE in terms of time and space averages. The quantum unique ergodicity problem (the term is due to Rudnick-Sarnak [**RS**]) is the following:

PROBLEM 4.2. *Suppose the geodesic flow g^t of (M, g) is ergodic on S^*M. Is the operator K in*

$$\langle A \rangle = \omega(A) + K$$

a compact operator? Equivalently is $\mathcal{Q} = \{d\mu_L\}$? In this case, $\sqrt{\Delta}$ is said to be QUE (quantum uniquely ergodic)

Compactness of K implies that $\langle K\varphi_k, \varphi_k\rangle \to 0$, hence $\langle A\varphi_k, \varphi_k\rangle \to \omega(A)$ along the entire sequence.

Rudnick-Sarnak conjectured that Δ of negatively curved manifolds are QUE, i.e. that for any orthonormal basis of eigenfunctions, the Liouville measure is the only quantum limit [**RS**].

4.5. Converse QE. So far we have not mentioned Theorem 4.1 (2). An interesting open problem is the extent to which (2) is actually necessary for the equivalence to classical ergodicity.

PROBLEM 4.3. *Suppose that $\sqrt{\Delta}$ is quantum ergodic in the sense that (1) holds in Theorem 4.1. What are the properties of the geodesic flow g^t. Is it ergodic (in the generic case)?*

In the larger class of Schrödinger operators, there is a simple example of a Hamiltonian system which is quantum ergodic but not classically ergodic: namely, a Schrödinger operator with a symmetric double well potential W. That is, W is a W shaped potential with two wells and a $\mathbb{Z}_2$ symmetry exchanging the wells. The low energy levels consist of two connected components interchanged by the symmetry, and hence the classical Hamiltonian flow is not ergodic. However, all eigenfunctions of the Schrödinger operator $-\frac{d^2}{dx^2} + W$ are either even or odd and thus have the same mass in both wells. It is easy to see that the quantum Hamiltonian is quantum ergodic.

Recently, B. Gutkin [**Gut**] has given a two dimensional example of a domain with boundary which is quantum ergodic but not classically ergodic and which is a two dimensional analogue of a double well potential. The domain is a so-called hippodrome (race-track) stadium. Similarly to the double well potential, there are two invariant sets interchanged by a $\mathbb{Z}_2$

symmetry. They correspond to the two orientations with which the race could occur. Hence the classical billiard flow on the domain is not ergodic. After dividing by the $\mathbb{Z}^2$ symmetry the hippodrome has ergodic billiards, hence by Theorem 4.1, the quotient domain is quantum ergodic. But the The eigenfunctions are again either even or odd. Hence the hippodrome is quantum ergodic but not classically ergodic.

Little is known about converse quantum ergodicity in the abscence of symmetry. It is known that if there exists an open set in S^*M filled by periodic orbits, then the Laplacian cannot be quantum ergodic (see [**MOZ**] for recent results and references). But it is not even known at this time whether KAM systems, which have Cantor-like invariant sets of positive measure, are not quantum ergodic. It is known that there exist a positive proportion of approximate eigenfunctions (quasi-modes) which localize on the invariant tori, but it has not been proved that a positive proportion of actual eigenfunctions have this localization property.

4.6. Quantum weak mixing. There are parallel results on quantizations of weak-mixing geodesic flows. We recall that the geodesic flow of (M, g) is weak mixing if the operator V^t has purely continuous spectrum on the orthogonal complement of the constant functions in $L^2(S^*M, d\mu_L)$.

THEOREM 4.4. [**Z8**] *The geodesic flow g^t of (M, g) is weak mixing if and only if the conditions (1)-(2) of Theorem 4.1 hold and additionally, for any $A \in \Psi^o(M)$,*

$$(\forall \epsilon)(\exists \delta) \limsup_{\lambda \to \infty} \frac{1}{N(\lambda)} \sum_{\substack{j \neq k: \lambda_j, \lambda_k \leq \lambda \\ |\lambda_j - \lambda_k - \tau| < \delta}} |(A\varphi_j, \varphi_k)|^2 < \epsilon \quad (\forall \tau \in \mathbb{R})$$

The restriction $j \neq k$ is of course redundant unless $\tau = 0$, in which case the statement coincides with quantum ergodicity. This result follows from the general asymptotic formula, valid for any compact Riemannian manifold (M, g), that

$$\begin{aligned} &\frac{1}{N(\lambda)} \sum_{i \neq j, \lambda_i, \lambda_j \leq \lambda} |\langle A\varphi_i, \varphi_j \rangle|^2 \left| \frac{\sin T(\lambda_i - \lambda_j - \tau)}{T(\lambda_i - \lambda_j - \tau)} \right|^2 \\ &\sim \| \frac{1}{2T} \int_{-T}^{T} e^{it\tau} V_t(\sigma_A) \|_2^2 - | \frac{\sin T\tau}{T\tau} |^2 \omega(A)^2. \end{aligned} \tag{56}$$

In the case of weak-mixing geodesic flows, the right hand side $\to 0$ as $T \to \infty$. As with diagonal sums, the sharper result is true where one averages over the short intervals $[\lambda, \lambda + 1]$.

Theorem 4.4 is based on expressing the spectral measures of the geodesic flow in terms of matrix elements. The main limit formula is:

$$\int_{\tau - \varepsilon}^{\tau + \varepsilon} d\mu_{\sigma_A} := \lim_{\lambda \to \infty} \frac{1}{N(\lambda)} \sum_{i,j:\ \lambda_j \leq \lambda,\ |\lambda_i - \lambda_j - \tau| < \varepsilon} |\langle A\varphi_i, \varphi_j \rangle|^2, \tag{57}$$

where $d\mu_{\sigma_A}$ is the spectral measure for the geodesic flow corresponding to the principal symbol of A, $\sigma_A \in C^\infty(S^*M, d\mu_L)$. Recall that the spectral measure of V^t corresponding to $f \in L^2$ is the measure $d\mu_f$ defined by

$$\langle V^t f, f\rangle_{L^2(S^*M)} = \int_{\mathbb{R}} e^{\mathrm{it}\,\tau} d\mu_f(\tau).$$

4.7. Evolution of Lagrangian states. In this section, we briefly review results on evolution of Lagrangian states and coherent states. We follow in particular the article of R. Schubert [**Schu3**].

A simple Lagrangian or WKB state has the form $\psi_\hbar(x) = a(\hbar, x)e^{\frac{i}{\hbar}S(x)}$ where $a(\hbar, x)$ is a semi-classical symbol $a \sim \sum_{j=0}^\infty \hbar^j a_j(x)$. The phase S generates the Lagrangian submanifold $(x, dS(x)) \subset T^*M$.

It is proved in Theorem 1 of [**Schu3**] that if g^t is Anosov and if Λ is transversal to the stable foliation W^s (except on a set of codimension one), then there exists $C, \tau > 0$ so that for every smooth density on Λ and every smooth function $a \in C^\infty(S^*M)$, the Lagrangian state ψ with symbol σ_ψ satisfies,

$$\left|\langle U^t\psi, AU^t\psi\rangle - \int_{S^*M}\sigma_A d\mu_L \int_\Lambda |\sigma_\psi|^2\right| \leq Che^{\Gamma|t|} + ce^{-t\tau}. \tag{58}$$

In order that the right side tends to zero as $\hbar \to 0, t \to \infty$ it is necessary and sufficient that

$$t \leq \frac{1-\epsilon}{\Gamma}|\log\hbar|.$$

5. Concentration of eigenfunctions around hyperbolic closed geodesics

As mentioned above, the quantum ergodicity Theorem 4.1 leaves open the possible existence of a sparse (zero density) subsequence of eigenfunctions which "weakly scar" on a hyperbolic orbit γ in the sense that its quantum limit ν_0 contains a non-trivial multiple of the periodic orbit measure $c\mu_\gamma$ as a non-zero ergodic component. The Anantharaman entropy bound shows that when (M, g) is Anosov, there cannot exist such a sequence of eigenfunctions (or even quasi-modes) which tend to μ_γ itself, but a quantum limit could have the form $c_1\mu_\gamma + c_2\mu_L$ for certain c_1, c_2 satisfying $c_1 + c_2 = 1$. The question we address in this section is the possible mass profile of such a scarring eigenfunction in a neighborhood of γ. More precisely, *how much mass does φ_λ have in a shrinking $\hbar^{1/2-\epsilon}$ tube around a hyperbolic closed geodesic?* This question will surface again in §11.21.

We first note that the existence of a quantum limit of the form $c_1\mu_\gamma + c_2\mu_L$ for (M, g) with Anosov geodesic flow is not so implausible. Analogous eigenfunction sequences do exist for the so-called quantum cat map [**FNB, FN**]. And exceptional sequences "scarring" on a certain 1-parameter family of periodic orbits exists for the Bunimovich stadium [**Has**]. At this time, there are no known examples of sequences of eigenfunctions or quasi-modes for (M, g) with ergodic geodesic flow that "weakly scar" along a hyperbolic closed orbit γ and no results prohibiting them.

For simplicity, assume that (M, g) is a Riemannian manifold of dimension 2 with a closed geodesic γ of length L. We assume γ is an embedded (non self-intersecting) curve. If φ_j is a Laplace eigenfunction, we define the mass profile of φ_j near γ to be the function

$$M(\varphi_j)(r) = \int_{d(x,\gamma)=r} |\varphi_j|^2 dS,$$

where $dS \wedge dr = dV$.

Before considering possible mass profiles of eigenfunctions scarring on hyperbolic closed geodesics, let us recall the opposite and much better known case of scarring of Gaussian beams along elliptic closed geodesics on surfaces [**Ra, Ra2, B.B**]. When γ is an elliptic closed geodesic, then there always exists a sequence of quasi-modes (Gaussian beams) which concentrates on γ. As the name suggests, Gaussian beams ψ_λ oscillate like $e^{i\lambda s}$ along γ and resemble Gaussians $\sqrt{\lambda} e^{-\lambda \langle A(s) y, y \rangle}$ (for some positive symmetric matrix $A(s)$) in the transverse direction with height $\lambda^{1/4}$ and concentrated in a $\frac{1}{\sqrt{\lambda}}$ tube around γ. Thus, the mass profile is a Gaussian probability measure with mean zero and variance $\lambda^{-1/2}$. The local model for such quasi-modes is that of a harmonic oscillator in the fibers of N_γ. One can construct the quasi-mode so that it is of infinite order. Of course, stable elliptic orbits do not exist when (M, g) has ergodic geodesic flow (by the KAM theorem).

Now consider hyperbolic closed orbits. It was pointed out by Duistermaat [**Dui**] (Section 1.5) that one cannot construct analogous quasi-modes associated to hyperbolic closed geodesics as Lagrangian states. The stable/unstable manifolds $\Lambda_\pm$ of γ, containing the geodesics which spiral in towards γ, are invariant Lagrangian submanifolds, but the only invariant half-density on the Lagrangians is the "delta"-density on the closed geodesic.

Further, there are apriori limitations on the degree to which eigenfunctions sequences can concentrate around hyperbolic closed geodesic on any (M, g).

THEOREM 5.1. [**BZ, Chr, CVP**] *Let (M, g) be a compact Riemannian manifold, and let γ be a hyperbolic closed geodesic. Let U be any tubular neighborhood of γ in M. Then for any eigenfunction φ_λ, there exists a constant C depending only on U such that*

$$\int_{M \setminus U} |\varphi_\lambda|^2 dV_g \geq \frac{C}{\log \lambda} ||\varphi_\lambda||_{L^2}^2.$$

More generally, let $A \in \Psi^0(M)$ be a pseudo-differential orbit whose symbol equals one in a neighborhood of γ in $S_g^* M$ and equals zero outside another neighborhood. Then for any eigenfunction φ_λ $||(I - A)\varphi_\lambda||_{L^2} \geq \frac{C}{\sqrt{\log \lambda}} ||\varphi_\lambda||_{L^2}$.

This allows scarring sequences along a hyperbolic orbit to occur, it just limits the rate at which the mass concentrates near γ. It implies that the mass profile of a sequence of eigenfunctions concentrating on a hyperbolic close geodesic has "long tails", i.e. there is a fairly large amount of mass far away from the geodesic, although sequences with the quantum limit μ_γ must tend to zero outside of any tube around the closed geodesic. Note that this result makes no dynamical hypotheses. It applies equally to (M, g) with integrable geodesic flow and to those with Anosov geodesic flow. To the author's knowledge, there do not exist more precise results in the Anosov case.

An obvious question at this point is whether there are any examples of (M, g), with any type of geodesic flow, possessing a sequence of eigenfunctions scarring on a closed hyperbolic orbit. The answer to this question is "yes". It is simple to see that such eigenfunctions exist in the opposite extreme of completely integrable systems, for instances surfaces of revolution like peanuts with hyperbolic waists. Examples include joint eigenfunctions of the square root of the Laplacian and rotation on surfaces of revolution with a hyperbolic waist. A truncated hyperbolic cylinder is another example studied in [**CVP**]. In this case, the joint spectrum fills out the image of the moment map $(p_\theta, |\xi|) : T^*M \to \mathbb{R}^2$, where $p_\theta(x, \xi) = \langle \xi, \frac{\partial}{\partial \theta} \rangle$ is the angular momentum. At critical distances to the axis of rotation, the lattitude circles are closed geodesics and the level set of the moment map becomes singular. If the surface is shaped like a peanut, the waist is a hyperbolic closed geodesic. Joint eigenfunctions whose joint eigenvalues are asymptotic to the singular levels of the moment always exist. The modes concentrate on the level sets of the moment map, and in fact they concentrate on the hyperbolic closed geodesic.

5.1. Mass concentration of special eigenfunctions on hyperbolic orbits in the quantum integrable case. The mass profile of scarring eigenfunctions near a hyperbolic in the completely integrable case is studied in [**CVP**] on tubes of fixed radius and in [**NV, TZ2**] on tubes of shrinking radius. Let $\gamma \subset S^*M$ be a closed hyperbolic geodesic of an (M, g) with completely integrable geodesic flow and for which Δ_g is quantum integrable (i.e. commutes with a maximal set of pseudo-differential operators; see [**TZ2**] for background). We then consider joint eigenfunctions Δ_g and of these operators. It is known (see e.g. [**TZ2**], Lemma 6) that there exists a special sequence of eigenfunctions concentrating on the momentum level set of γ. We will call them (in these notes) the *γ-sequence.*

Assume for simplicity that the moment level set of γ just consists of the orbit together with its stable/unstable manifolds. Then it is proved in [**TZ2**] that the mass of φ_μ in the shrinking tube of radius h^δ around γ with $\delta < \frac{1}{2}$ is $\simeq (1 - 2\delta)$ (see also [**NV**] for a closely related result in two dimensions). Thus, the mass profile of such scarring integrable eigenfunctions only differs by the numerical factor $(1 - 2\delta)$ from the mass profile of Gaussian beams.

The difference is that the "tails" in the hyperbolic case are longer. Also the peak is logarithmically smaller than in the elliptic case (a somewhat weaker statement is proved in [**TZ2**]).

Let us state the result precisely and briefly sketch the argument. It makes an interesting comparison to the situation discussed later on of possible scarring in the Anosov case.

We denote by $\pi : S^*M \to M$ the standard projection and let $\pi(\gamma)$ be the image of γ in M. We denote by $T_\epsilon(\pi(\Lambda))$ the tube of radius ϵ around $\pi(\Lambda)$. For $0 < \delta < 1/2$, we introduce a cutoff $\chi_1^\delta(x;\hbar) \in C_0^\infty(M)$ with $0 \le \chi_1^\delta \le 1$, satisfying

- (i) supp $\chi_1^\delta \subset T_{\hbar^\delta}(\pi(\gamma))$
- (ii) $\chi_1^\delta = 1$ on $T_{3/4\hbar^\delta}(\pi(\gamma))$.

THEOREM 5.2. *Let γ be a hyperbolic closed orbit in (M,g) with quantum integrable Δ_g, and let $\{\varphi_\mu\}$ be an L^2 normalized γ-sequence of joint eigenfunctions Then for any $0 \le \delta < 1/2$, $\lim_{\hbar\to 0}(Op_\hbar(\chi_1^\delta)\varphi_\mu,\varphi_\mu) \ge (1-2\delta)$.*

5.1.1. *Outline of proof.* For simplicity we assume $\dim M = 2$. Let $\chi_2^\delta(x,\xi;\hbar) \in C_0^\infty(T^*M;[0,1])$ be a second cutoff supported in a radius $\hbar^\delta$ tube, $\Omega(\hbar)$, around γ with $\Omega(\hbar) \subset supp\chi_1^\delta$ and such that $\chi_1^\delta = 1$ on $supp\chi_2^\delta$. Thus, $\chi_1^\delta(x,\xi) \ge \chi_2^\delta(x,\xi)$, for any $(x,\xi) \in T^*M$. By the Garding inequality, there exists a constant $C_1 > 0$ such that:

$$(Op_\hbar(\chi_1^\delta)\varphi_\mu,\varphi_\mu)) \ge (Op_\hbar(\chi_2^\delta)\varphi_\mu,\varphi_\mu) - C_1\hbar^{1-2\delta}. \tag{59}$$

We now conjugate the right side to the model setting of $S^1 \times \mathbb{R}^1$, i.e. the normal bundle N_γ to γ. The conjugation is done by $\hbar$ Fourier integral operators and is known as conjugation to quantum Birkhoff normal form. In the model space, the conjugate of Δ_g is a function of $D_s = \frac{\partial}{i\partial s}$ along S^1 and the dilation operator $\hat{I}^h := \hbar(D_y y + yD_y)$ along $\mathbb{R}$. By Egorov's theorem

$$(Op_\hbar(\chi_2^\delta)\varphi_\mu,\varphi_\mu) = |c(\hbar)|^2(Op_\hbar(\chi_2^\delta\circ\kappa)u_\mu,u_\mu) - C_3\hbar^{1-2\delta} \tag{60}$$

where $u_\mu(y,s;\hbar)$ is the model joint eigenfunction of $D_s, \hat{I}^h$, and $c(\hbar)$ is a normalizing constant. This reduces our problem to estimating the explicit matrix elements $(Op_\hbar(\chi_2^\delta\circ\kappa)u_\mu,u_\mu)$ of the special eigenfunctions in the model setting. The operator $\hat{I}^h$ has a continuous spectrum with generalized eigenfunctions $y^{-1/2+i\lambda/\hbar}$. The eigenfunctions on the "singular" level γ correspond to $\lambda \sim E\hbar$. A caluclation shows that the mass in the model setting is given by

$$M_h = \frac{1}{\log\hbar}\left(\int_0^\infty \chi(\hbar\xi/\hbar^\delta)\left|\int_0^\infty e^{-ix}x^{-1/2+i\lambda/\hbar}\chi(x/\hbar^\delta\xi)dx\right|^2\frac{d\xi}{\xi}\right). \tag{61}$$

Analysis of (61) shows that the right side tends to $1-2\delta$ as $\hbar\to 0$ if $\lambda\sim E\hbar$ (see §5.3 for a detailed discussion).

5.2. Comparison to Anosov case. This large mass profile may be a special feature of integrable systems, reflecting the fact that the stable and unstable manifolds of the hyperbolic closed orbit coincide. A heuristic picture of the mass concentration in this case is as follows: Since the eigenfunction is a stationary state, its mass must be asymptotically invariant under the geodesic flow. Since the flow compresses things exponentially in the stable direction and expands things in the unstable direction, the mass can only concentrate on the fixed closed geodesic and on the unstable manifold W^u. But W^u returns to γ as the stable manifold W^u (like a figure 8). Hence, the only invariant measure is the one supported on the closed geodesic and the the mass can only concentrate there. Although it does not seem to have been proved in detail yet, it is very plausible that the mass concentration in the integrable case provides an upper bound for any (M, g), i.e. it has "extremal" mass concentration.

In the Anosov case, the stable and unstable manifolds of γ are transverse, so the dynamical picture is completely different. First, there is no obvious mechanism as in the integrable case forcing mass of any sequence of eigenfunctions to concentrate on the Lagrangian manifold formed by γ and W^u in the Anosov case. If mass did concentrate around γ, it would still be forced to concentrate on γ and on W^u, but W^u becomes dense in S^*M. Hence some of the mass must spread out uniformly over S^*M and is lost from a neighborhood of γ. This makes it plausible that one does not get mass concentration for eigenfunctions around hyperbolic closed orbits of Anosov systems.

Yet, in the "cat map" analogue, there do exist scarring eigenfunctions [**FNB, FN**] for a special sparse sequence of Planck constants. The multiplicities of the eigenvalues of the cat map for this sequence saturate the bound $\hbar^{-1}/|\log \hbar|$, and therefore one can build up eigenfunctions with very unexpected properties. There is a surprising quantum mechanism forcing concentration of special modes at hyperbolic fixed points which was discovered by Faure-Nonnenmacher-de Bièvre. The eigenfunction amplitude spreads out along $\hbar^{-1/2}$ segments of W^u. These segments are $\hbar^{1/2}$ dense and so there is interference between the amplitudes on close pieces of W^u. The interference is constructive along W^s and the mass builds up there and then as in the integrable case gets recycled back to the hyperbolic orbit.

It is not known whether this phenomenon occurs in the Riemannian setting. It is presumably related to the existence of sparse subsequences of eigenvalues with the same large multiplicities.

5.3. Details on M_h. The proof that (61) tends to $1 - 2\delta$ as $\hbar \to 0$ if $\lambda \sim E\hbar$ is rather sketchy in [**TZ2**]. In dimension one, the proof in a model case was given in [**NV**] (we thank S. Nonnenmacher for a clarifying discussion on this point). We supply the details in the general case here as an addendum to [**TZ2**], to which we refer for further background.

To estimate M_h, we assume $\lambda \sim E\hbar$ and let $\chi \in C_0^\infty(\mathbb{R};[0,1])$ with $\chi(x) = 1$ for $|x| \leq 1$ and $\chi(x) = 0$ for $|x| \geq 2$. Note that we divide by $|\log \hbar|$ in the model distribution $u_h(x) = \frac{1}{\sqrt{|\log \hbar|}} x^{-1/2+iE/\hbar} Y(x)$, so that $\|Op_\hbar(\chi)u_h\|_{L^2} \sim 1$. To estimate the mass on shrinking tubes of size $\hbar^\delta$ we write

$$
\begin{aligned}
(62) \quad M_h &= |\log \hbar|^{-1} \int_0^\infty \chi(\hbar^{1-\delta}\xi) \left| \int_0^\infty e^{-ix} x^{-1/2+i\lambda/\hbar} \chi(x/\hbar^\delta \xi) dx \right|^2 \frac{d\xi}{\xi} \\
&= |\log \hbar|^{-1} \int_0^\infty \chi(\hbar^{1-\delta}\xi) \left| \int_0^\infty e^{-ix} x^{-1/2+i\lambda/\hbar} \chi(x/\hbar^\delta \xi) dx \right|^2 \frac{d\xi}{\xi} \\
&= |\log \hbar|^{-1} \int_0^{\hbar^{\delta-1}} \frac{d\xi}{\xi} \left| \int_0^\infty e^{-ix} x^{-1/2+i\lambda/\hbar} \chi(x/\hbar^\delta \xi) dx \right|^2 \\
&\quad + \mathcal{O}(|\log \hbar|^{-1}) \ (*)
\end{aligned}
$$

The last step follows since

$$
\begin{aligned}
&|\log \hbar|^{-1} \int_{\hbar^{\delta-1}}^{2\hbar^{\delta-1}} \frac{d\xi}{\xi} \left| \int_0^\infty e^{-ix} x^{-1/2+i\lambda/\hbar} \chi(x/\hbar^\delta \xi) dx \right|^2 \\
&= |\log \hbar|^{-1} \int_{\hbar^{\delta-1}}^{2\hbar^{\delta-1}} \frac{d\xi}{\xi} \left| \int_1^\infty e^{-ix} x^{-1/2+i\lambda/\hbar} \chi(x/\hbar^\delta \xi) dx + \mathcal{O}(1) \right|^2 \\
&= \mathcal{O}(|\log \hbar|^{-1}).
\end{aligned}
$$

The last estimate follows by integration by parts, since when $\xi \geq \hbar^{\delta-1}$ we have $\hbar^\delta \xi \geq \hbar^{2\delta-1}$ and we assume that $2\delta - 1 < 0$ so $D_x(\chi(x/\hbar^\delta \xi)) = \mathcal{O}(\hbar^{-\delta}\xi^{-1}) \to 0$. Also, $|\Gamma(1/2 + i\lambda/\hbar)|^2 = |\Gamma(1/2 + iE)|^2 = \frac{\pi}{\cosh(\pi E)} = O(1)$. To simplify $(*)$ we first make a change of variables $\xi \mapsto \hbar^\delta \xi$ and get

$$
M_h = |\log \hbar|^{-1} \int_0^{\hbar^{2\delta-1}} \frac{d\eta}{\eta} \left| \int_0^\infty e^{-ix} x^{-1/2+i\lambda/\hbar} \chi(x/\eta) dx \right|^2 + \mathcal{O}(|\log \hbar|^{-1}) \ (**)
$$

Next, we get rid of the interval $0 \leq \eta \leq 1$ by observing that when $\eta \in [0,1]$,

$$
\left| \int_0^\infty e^{-ix} x^{-1/2+i\lambda/\hbar} \chi(x/\eta) dx \right| \leq \int_0^{2\eta} x^{-1/2} dx = \mathcal{O}(\eta^{1/2})
$$

and so,

$$
\begin{aligned}
&|\log \hbar|^{-1} \int_0^1 \frac{d\eta}{\eta} \left| \int_0^\infty e^{-ix} x^{-1/2+i\lambda/\hbar} \chi(x/\eta) dx \right|^2 \ll |\log \hbar|^{-1} \int_0^1 \eta^{-1} \eta d\eta \\
&\ll |\log \hbar|^{-1}.
\end{aligned}
$$

Thus,

(63)

$$
M_h = |\log \hbar|^{-1} \int_1^{\hbar^{2\delta-1}} \frac{d\eta}{\eta} \left| \int_0^\infty e^{-ix} x^{-1/2+i\lambda/\hbar} \chi(x/\eta) dx \right|^2 + \mathcal{O}(|\log \hbar|^{-1}).
$$

Next, one gets rid of the cutoff $\chi(x/\eta)$ by integrating by parts. When $\eta \geq 1$,

$$\int_\eta^\infty e^{-ix} x^{-1/2+i\lambda/\hbar} \chi(x/\eta) dx = \int_\eta^\infty D_x(e^{-ix}) x^{-1/2+i\lambda/\hbar} \chi(x/\eta) dx$$
$$= \mathcal{O}(\eta^{-1/2}).$$

So the $[\eta, \infty]$-range of integration in x gives a contribution to M_h that is bounded by

$$C |\log \hbar|^{-1} \int_1^{\hbar^{2\delta-1}} \frac{d\eta}{\eta} \eta^{-1} \ll |\log \hbar|^{-1}.$$

The end result is the formula

$$M_h = |\log \hbar|^{-1} \int_1^{\hbar^{2\delta-1}} \frac{d\eta}{\eta} \left| \int_0^\eta e^{-ix} x^{-1/2+i\lambda/\hbar} dx \right|^2 + \mathcal{O}(|\log \hbar|^{-1}). \tag{64}$$

By contour deformation, for $\eta \geq 1$,

$$\left| \int_0^\eta e^{-ix} x^{-1/2+i\lambda/\hbar} dx \right|^2 = \left| \Gamma(1/2 + i\lambda/\hbar) + \mathcal{O}(\eta^{-1/2}) \right|^2 \quad (***)$$

Substitution into (64) gives

$$M_h = |\log \hbar|^{-1} |\Gamma(1/2 + i\lambda/\hbar)|^2 \int_1^{\hbar^{2\delta-1}} \frac{d\eta}{\eta} + \mathcal{O}(|\log \hbar|^{-1}). \tag{65}$$

This last step follows since the $\mathcal{O}(\eta^{-1/2})$-terms in $(***)$ give a contribution to $M_\hbar$ that is $\ll |\log \hbar|^{-1} \int_1^{\hbar^{2\delta-1}} \frac{d\eta}{\eta} \eta^{-1/2} = \mathcal{O}(|\log \hbar|^{-1})$.

The final formula follows immediately from (65). With $\lambda \sim E\hbar$, one gets

$$M_h \sim |\Gamma(1/2 + iE)|^2 (1 - 2\delta) \tag{66}$$

Since $|\Gamma(1/2 + iE)|^2 = \frac{\pi}{\cosh \pi E}$, which is bounded away from zero, it follows that (61) tends to $1 - 2\delta$ as $\hbar \to 0$ □

6. Boundary quantum ergodicity and quantum ergodic restriction

In this section, we briefly discuss quantum ergodic restriction theorems. The general question is whether restrictions of quantum ergodic eigenfunctions to hypersurfaces (or microlocally, to cross-sections of the geodesic flow) are quantum ergodic on the hypersurfaces. We only consider here the case where the hypersurface is the boundary of a domain with boundary, which was studied in [**GL, HZ, Bu**]. More general quantum ergodic restriction theorems are given in [**TZ3**]. The boundary results play an important role in the recent scarring results of A. Hassell for eigenfunctions on the stadium.

We thus consider the boundary values u_j^b of interior eigenfunctions

$$\begin{cases} \Delta\, u_j = \lambda_j^2\, u_j \ \text{ in } \Omega, & \langle u_j, u_k \rangle_{L^2(\Omega)} = \delta_{jk}, \\ \\ B u_j|_Y = 0, \quad Y = \partial\Omega \end{cases}$$

of the Euclidean Laplacian Δ on a compact piecewise smooth domain $\Omega \subset \mathbb{R}^n$ and with classically ergodic billiard map $\beta : B^*Y \to B^*Y$, where $Y = \partial\Omega$. Here A_h is a zeroth order semiclassical pseudodifferential operator on Y. The relevant notion of boundary values (i.e. Cauchy data) u_j^b depends on the boundary condition. We only consider the boundary conditions

$$Bu = \begin{cases} u|_Y, & \text{Dirichlet} \\ \\ \partial_\nu u|Y, & \text{Neumann} \end{cases} \tag{67}$$

Let Δ_B denote the positive Laplacian on Ω with boundary conditions $Bu = 0$. Then Δ_B has discrete spectrum $0 < \lambda_1 < \lambda_2 \leq \cdots \to \infty$, where we repeat each eigenvalue according to its multiplicity, and for each λ_j we may choose an L^2 normalized eigenfunction u_j.

The algebra of observables in the boundary setting is the algebra $\Psi_h^0(Y)$ of zeroth order semiclassical pseudodifferential operators on Y, depending on the parameter $h \in [0, h_0]$. We denote the symbol of $A = A_h \in \Psi_h^0(Y)$ by $a = a(y, \eta, h)$. Thus $a(y, \eta) = a(y, \eta, 0)$ is a smooth function on T^*Y.

To each boundary condition B corresponds

- A specific notion of boundary value u_j^b of the eigenfunctions u_j. We denote the L^2-normalized boundary values by $\hat{u}_j^b = u_j^b / ||u_j^b||$.
- A specific measure $d\mu_B$ on $B^*(Y)$.
- A specific state ω_B on the space $\Psi_h^0(Y)$ of semiclassical pseudodifferential operators of order zero defined by

$$\omega_B(A) = \frac{4}{\mathrm{vol}(S^{n-1})\mathrm{vol}(\Omega)} \int_{B^*Y} a(y, \eta) d\mu_B. \tag{68}$$

Here is a table of the relevant boundary value notions. In the table, $d\sigma$ is the natural symplectic volume measure on B^*Y. We also define the function $\gamma(q)$ on B^*Y by

$$\gamma(q) = \sqrt{1 - |\eta|^2}, \quad q = (y, \eta). \tag{69}$$

Boundary Values			
B	Bu	u^b	$d\mu_B$
Dirichlet	$u\|_Y$	$\lambda^{-1}\, \partial_\nu u\|_Y$	$\gamma(q) d\sigma$
Neumann	$\partial_\nu u\|_Y$	$u\|_Y$	$\gamma(q)^{-1} d\sigma$

The limit states are determined by dictated by the local Weyl law for the boundary condition B.

LEMMA 6.1. *Let A_h be either the identity operator on Y or a zeroth order semiclassical operator on Y with kernel supported away from the singular set. Then for any of the above boundary conditions B, we have:*

$$\begin{aligned}
&\lim_{\lambda\to\infty}\frac{1}{N(\lambda)}\sum_{\lambda_j\leq\lambda}\langle A_{h_j}u_j^b,u_j^b\rangle=\omega_B(A),\quad B=\textit{Neumann},\\
&\lim_{\lambda\to\infty}\frac{1}{N(\lambda)}\sum_{\lambda_j\leq\lambda}\langle A_{h_j}u_j^b,u_j^b\rangle=\omega_B(A),\quad B=\textit{Dirichlet}.
\end{aligned}\tag{70}$$

The boundary quantum ergodic theorem is:

THEOREM 6.2. **[HZ, GL, Bu]** *Let $\Omega\subset\mathbb{R}^n$ be a bounded piecewise smooth manifold with ergodic billiard map. Let $\{u_j^b\}$ be the boundary values of the eigenfunctions $\{u_j\}$ of Δ_B on $L^2(\Omega)$ in the sense of the table above. Let A_h be a semiclassical operator of order zero on Y. Then there is a subset S of the positive integers, of density one, such that*

$$\begin{aligned}
&\lim_{j\to\infty,j\in S}\langle A_{h_j}u_j^b,u_j^b\rangle=\omega_B(A),\quad B=\textit{Neumann},\\
&\lim_{j\to\infty,j\in S}\langle A_{h_j}u_j^b,u_j^b\rangle=\omega_B(A),\quad B=\textit{Dirichlet},
\end{aligned}\tag{71}$$

where $h_j=\lambda_j^{-1}$ and ω_B is as in (68).

In the case $A=I$ and for the Neumann boundary condition, we have

$$\lim_{j\to\infty,j\in S}\|u_j^b\|^2_{L^2(Y)}=\frac{2\mathrm{vol}(Y)}{\mathrm{vol}(\Omega)},$$

while for the Dirichlet boundary condition,

$$\lim_{j\to\infty,j\in S}\|u_j^b\|^2_{L^2(Y)}=\frac{2\mathrm{vol}(Y)}{n\mathrm{vol}(\Omega)}.$$

6.1. Sketch of proof. The fact that the quantum limit state ω_B and the corresponding measure $d\mu_B$ are not the natural symplectic volume measure $d\sigma$ on B^*Y is due to the fact that the quantum dynamics is defined by an endomorphism rather than an automorphism of the observable algebra. In the Neumann case, the dynamics are generated by the operator F_h on Y with kernel

$$\begin{aligned}
&F_h(y,y')=2\frac{\partial}{\partial\nu_y}G_0(y,y',h^{-1}),\quad y\neq y'\in Y,\ \text{where}\\
&G_0(y,y',\lambda)=\frac{i}{4}\lambda^{d-2}(2\pi\lambda|z-z'|)^{-(d-2)/2}\,\mathrm{Ha}^{(1)}_{d/2-1}(\lambda|z-z'|)
\end{aligned}\tag{72}$$

is the free outgoing Green function on $\mathbb{R}^d$. These are (almost) semi-classical Fourier integral operators whose phase functions, the boundary distance function $d_b(y,y')=|y-y'$ $(y,y'\in Y)$ generates the billiard map. It is well known that this operator leaves the boundary values of Neumann eigenfunctions u_j^b invariant:

$$F_{h_j} u_j^b = u_j^b, \quad j = 1, 2, \ldots \tag{73}$$

It follows that the states

$$\rho_j(A) := \langle A_{h_j} u_j^b, u_j^b \rangle \tag{74}$$

are invariant for F_{h_j}. The family $\{F_h\}$ defines a semiclassical Fourier integral operator associated to the billiard map β (for convex Ω), plus some negligible terms. The quantum dynamics on $\Psi_h^0(Y)$ is thus generated by the conjugation

$$\alpha_{h_j}(A_{h_j}) = F_{h_j}^* \, A_{h_j} \, F_{h_j} \tag{75}$$

This is analogous to the interior dynamics generated by the wave group U_B^t with the boundary conditions, but unlike U_B^t, F_h is not unitary or even normal. Indeed, the zeroth order part of $F_h^* F_h$ is a semiclassical pseudodifferential operator with a non-constant symbol.

The Egorov type result for the operator F_h is as follows: Let β denote the billiard map on B^*Y^o and let $A_h = \mathrm{Op}(a_h)$ be a zeroth order operator whose symbol $a(y, \eta, 0)$. Let γ be given by (69). Then

$$F_h^* A_h F_h = \tilde{A}_h + S_h,$$

where $\tilde{A}_h$ is a zeroth order pseudodifferential operator and $\|S_h\|_{L^2 \to L^2} \leq Ch$. The symbol of $\tilde{A}_h$ is

$$\tilde{a} = \begin{cases} \gamma(q)\gamma(\beta(q))^{-1} a(\beta(q)), & q \in B^*Y \\ 0, & q \notin B^*Y. \end{cases} \tag{76}$$

This is a rigorous version of the statement that F_h quantizes the billiard ball map. This Egorov theorem is relevant to the Neumann boundary problem. In the Dirichlet case, the relevant operator is F_h^*.

The unusual transformation law of the symbol reflects the fact that (75) is not an automorphism. In the spectral theory of dynamical systems, one studies the dynamics of the billiard map β on B^*Y through the associated "Koopman" operator

$$\mathcal{U} : L^2(B^*Y, d\sigma) \to L^2(B^*Y, d\sigma), \;\; \mathcal{U}f(\zeta) = f(\beta(\zeta)).$$

Here, $d\sigma$ denotes the usual β-invariant symplectic volume measure on B^*Y. From the symplectic invariance it follows that $\mathcal{U}$ is a unitary operator. When β is ergodic, the unique invariant L^2-normalized eigenfunction is a constant c.

However, Egorov's theorem (76) in the boundary reduction involves the positive function $\gamma \in C^\infty(B^*Y)$, and the relevant Koopman operator is

$$Tf(\zeta) = \frac{\gamma(\zeta)}{(\gamma(\beta(\zeta)))} f(\beta(\zeta)).$$

Then T is not unitary on $L^2(B^*Y, d\sigma)$. Instead one has:

- (i) The unique positive T-invariant L^1 function is given by γ.
- (ii) T is unitary relative to the inner product $\langle\langle , \rangle\rangle$ on B^*Y defined by the measure $d\nu = \gamma^{-2} d\sigma$.
- (iii) When β is ergodic, the orthogonal projection P onto the invariant L^2-eigenvectors has the form
$$P(f) = \frac{\langle\langle f, \gamma \rangle\rangle}{\langle\langle \gamma, \gamma \rangle\rangle} \gamma = \frac{1}{\text{vol}(B^*Y)} [\int_{B^*Y} f \gamma^{-1} d\sigma]\, \gamma = c\omega_{\text{Neu}}(f)\, \gamma$$
where c is as in (117).

The proof of Theorem 6.2 then runs along similar lines to that of Theorem 4.1 but adjusted to the fact that that T is not unitary.

7. Hassell's scarring result for stadia

This section is an exposition of Hassell's scarring result for the Bunimovich stadium. We follow [**Has**] and [**Z7**].

A stadium is a domain $X = R \cup W \subset \mathbb{R}^2$ which is formed by a rectangle $R = [-\alpha, \alpha]_x \times [-\beta, \beta]_y$ and where $W = W_{-\beta} \cup W_\beta$ are half-discs of radius β attached at either end. We fix the height $\beta = \pi/2$ and let $\alpha = t\beta$ with $t \in [1, 2]$. The resulting stadium is denoted X_t.

It has long been suspected that there exist exceptional sequences of eigenfunctions of X which have a singular concentration on the set of "bouncing ball" orbits of R. These are the vertical orbits in the central rectangle that repeatedly bounce orthogonally against the flat part of the boundary. The unit tangent vectors to the orbits define an invariant Lagrangian submanifold with boundary $\Lambda \subset S^*X$. It is easy to construct approximate eigenfunctions which concentrate microlocally on this Lagrangian submanifold. Namely, let $\chi(x)$ be a smooth cutoff supported in the central rectangle and form $v_n = \chi(x) \sin ny$. Then for any pseudo-differential operator A properly supported in X,
$$\langle A v_n, v_n \rangle \to \int_\Lambda \sigma_A \chi d\nu$$
where $d\nu$ is the unique normalized invariant measure on Λ.

Numerical studies suggested that there also existed genuine eigenfunctions with the same limit. Recently, A. Hassell has proved this to be correct for almost all stadia.

THEOREM 7.1. *The Laplacian on X_t is not QUE for almost every $t \in$* $[1, 2]$.

We now sketch the proof and develop related ideas on quantum unique ergodicty. The main idea is that the existence of the scarring bouncing ball quasi-modes implies that either

- There exist actual modes with a similar scarring property, or

- The spectrum has exceptional clustering around the bouncing ball quasi-eigenvalues n^2.

Hassell then proves that the second alternative cannot occur for most stadia. We now explain the ideas in more detail.

We first recall that a quasi-mode $\{\psi_k\}$ is a sequence of L^2-normalized functions which solve

$$||(\Delta - \mu_k^2)\psi_k||_{L^2} = O(1),$$

for a sequence of quasi-eigenvalues μ_k^2. By the spectral theorem it follows that there must exist true eigenvalues in the interval $[\mu_k^2 - K, \mu_k^2 + K]$ for some $K > 0$. Moreover, if $\tilde{E}_{k,K}$ denotes the spectral projection for Δ corresponding to this interval, then

$$||\tilde{E}_{k,K}\psi_k - \psi_k||_{L^2} = O(K^{-1}).$$

To maintain consistency with (6), i.e. with our use of frequencies μ_k rather than energies μ_k^2, we re-phrase this in terms of the projection $E_{k,K}$ for $\sqrt{\Delta}$ in the interval $[\sqrt{\mu_k^2 - K}, \sqrt{\mu_k^2 + K}]$. For fixed K, this latter interval has width $\frac{K}{\mu_k}$.

Given a quasimode $\{\psi_k\}$, the question arises of how many true eigenfunctions it takes to build the quasi-mode up to a small error.

DEFINITION: We say that a quasimode $\{\psi_k\}$ of order 0 with $||\psi_k||_{L^2} = 1$ has $n(k)$ essential frequencies if

$$\psi_k = \sum_{j=1}^{n(k)} c_{kj}\varphi_j + \eta_k, \quad ||\eta_k||_{L^2} = o(1). \tag{77}$$

To be a quasi-mode of order zero, the frequencies λ_j of the φ_j must come from an interval $[\mu_k - \frac{K}{\mu_k}, \mu_k + \frac{K}{\mu_k}]$. Hence the number of essential frequencies is bounded above by the number $n(k) \leq N(k, \frac{K}{k})$ of eigenvalues in the interval. Weyl's law for $\sqrt{\Delta}$ allows considerable clustering and only gives $N(k, \frac{K}{k}) = o(k)$ in the case where periodic orbits have measure zero. For instance, the quasi-eigenvalue might be a true eigenvalue with multiplicity saturating the Weyl bound. But a typical interval has a uniformly bounded number of Δ-eigenvalues in dimension 2 or equivalently a frequency interval of with $O(\frac{1}{\mu_k})$ has a uniformly bounded number of frequencies. The dichotomy above reflects the dichotomy as to whether exceptional clustering of eigenvalues occurs around the quasi-eigenvalues n^2 of Δ or whether there is a uniform bound on $N(k, \delta)$.

PROPOSITION 7.2. *If there exists a quasi-mode $\{\psi_k\}$ of order 0 for Δ with the properties:*

- *(i) $n(k) \leq C$, $\forall\, k$;*

- *(ii)* $\langle A\psi_k, \psi_k\rangle \to \int_{S^*M} \sigma_A d\mu$ *where* $d\mu \neq d\mu_L$.

Then Δ *is not QUE.*

The proof is based on the following lemma pertaining to near off-diagonal Wigner distributions. It gives an "everywhere" version of the off-diagonal part of Theorem 4.1 (2).

LEMMA 7.3. *Suppose that* g^t *is ergodic and* Δ *is QUE. Suppose that* $\{(\lambda_{i_r}, \lambda_{j_r}),\ i_r \neq j_r\}$ *is a sequence of pairs of eigenvalues of* $\sqrt{\Delta}$ *such that* $\lambda_{i_r} - \lambda_{j_r} \to 0$ *as* $r \to \infty$. *Then* $dW_{i_r,j_r} \to 0$.

PROOF. Let $\{\lambda_i, \lambda_j\}$ be any sequence of pairs with the gap $\lambda_i - \lambda_j \to 0$. Then by Egorov's theorem, any weak* limit $d\nu$ of the sequence $\{dW_{i,j}\}$ is a measure invariant under the geodesic flow. The weak limit is defined by the property that

$$\langle A^* A\varphi_i, \varphi_j\rangle \to \int_{S^*M} |\sigma_A|^2 d\nu. \tag{78}$$

If the eigenfunctions are real, then $d\nu$ is a real (signed) measure.

We now observe that any such weak* limit must be a constant multiple of Liouville measure $d\mu_L$. Indeed, we first have:

$$|\langle A^* A\varphi_i, \varphi_j\rangle| \leq |\langle A^* A\varphi_i, \varphi_i\rangle|^{1/2}\ |\langle A^* A\varphi_j, \varphi_j\rangle|^{1/2}. \tag{79}$$

Taking the limit along the sequence of pairs, we obtain

$$|\int_{S^*M} |\sigma_A|^2 d\nu| \leq \int_{S^*M} |\sigma_A|^2 d\mu_L. \tag{80}$$

It follows that $d\nu << d\mu_L$ (absolutely continuous). But $d\mu_L$ is an ergodic measure, so if $d\nu = f d\mu_L$ is an invariant measure with $f \in L^1(d\mu_L)$, then f is constant. Thus,

$$d\nu = C d\mu_L, \quad \text{for some constant } C. \tag{81}$$

We now observe that $C = 0$ if $\varphi_i \perp \varphi_j$ (i.e. if $i \neq j$). This follows if we substitute $A = I$ in (78), use orthogonality and (81).

We now complete the proof of the Proposition by arguing by contradiction. The frequencies must come from a shrinking frequency interval, so the hypothesis of the Proposition is satisfied. If Δ were QUE, we would have (in the notation of (77):

$$\begin{aligned}\langle A\psi_k, \psi_k\rangle &= \textstyle\sum_{i,j=1}^{n(k)} c_{kj} c_{ki} \langle A\varphi_i, \varphi_j\rangle + o(1)\\ &= \textstyle\sum_{j=1}^{n(k)} c_{kj}^2 \langle A\varphi_j, \varphi_j\rangle + \sum_{i\neq j=1}^{n(k)} c_{kj} c_{ki} \langle A\varphi_i, \varphi_j\rangle + o(1)\\ &= \textstyle\int_{S^*M} \sigma_A d\mu_L + o(1),\end{aligned}$$

by Proposition 7.3. This contradicts (ii). In the last line we used $\sum_{j=1}^{n(k)} |c_{kj}|^2 = 1 + o(1)$, since $||\psi_k||_{L^2} = 1$.

QED

7.1. Proof of Hassell's scarring result. We apply and develop this reasoning in the case of the stadium. The quasi-eigenvalues of the Bunimovich stadium corresponding to bouncing ball quasi-modes are n^2 independently of the diameter t of the inner rectangle.

By the above, it suffices to show that that there exists a sequence $n_j \to \infty$ and a constant M (independent of j) so that there exist $\leq M$ eigenvalues of Δ in $[n_j^2 - K, n_j^2 + K]$. An somewhat different argument is given in [**Has**] in this case: For each n_j there exists a normalized eigenfunction u_{k_j} so that $\langle u_{k_j}, v_{k_j} \rangle \geq \sqrt{\frac{3}{4}M}$. It suffices to choose the eigenfunction with eigenvalue in the interval with the largest component in the direction of v_{k_j}. There exists one since

$$||\tilde{E}_{[n^2-K,n^2+K}v_n|| \geq \frac{3}{4}.$$

The sequence $\{u_{n_k}\}$ cannot be Liouville distributed. Indeed, for any $\epsilon > 0$, let A be a self-adjoint semi-classical pseudo-differential operator properly supported in the rectangle so that $\sigma_A \leq 1$ and so that $||(Id - A)v_n|| \leq \epsilon$ for large enough n. Then

$$\begin{aligned}\langle A^2 u_{k_j}, u_{k_j} \rangle &= ||A u_{k_j}||^2 \geq \left|\langle A u_{k_j}, v_{k_j} \rangle\right|^2 \\ &= \left|\langle u_{k_j}, A v_{k_j} \rangle\right|^2 \geq \left(|\langle u_{k_j}, v_{k_j} \rangle| - \epsilon\right)^2 \geq \left(\sqrt{\frac{3}{4}M} - \epsilon\right)^2.\end{aligned}$$

Choose a sequence of operators A such that $||(Id - A)v_n|| \to 0$ and so that the support of σ_A shrinks to the set of bouncing ball covectors. Then the mass of any quantum limit of $\{u_{n_k}\}$ must have mass $\geq \frac{3}{4}M$ on Λ.

Thus, the main point is to eliminate the possibility of exceptional clustering of eigenvalues around the quasi-eigenvalues. In fact, no reason is known why no exceptional clustering should occur. Hassell's idea is that it can however only occur for a measure zero set of diameters of the inner rectangle. The proof is based on Hadamard's variational formula for the variation of Dirichlet or Neumann eigenvalues under a variation of a domain. In the case at hand, the stadium is varied by horizontally (but not vertically) expanding the inner rectangle. In the simplest case of Dirichlet boundary conditions, the eigenvalues are forced to decrease as the rectangle is expanded. The QUE hypothesis forces them to decrease at a uniform rate. But then they can only rarely cluster at the fixed quasi-eigenvalues n^2. If this ever happened, the cluster would move left of n^2 and there would not be time for a new cluster to arrive.

Here is a more detailed sketch. Under the variation of X_t with infinitesimal variation vector field ρ_t, Hadamard's variational formula gives,

$$\frac{dE_j(t)}{dt} = \int_{\partial X_t} \rho_t(s)(\partial_n u_j(t)(s))^2 ds.$$

Then

$$E_j^{-1} \frac{d}{dt} E_j(t) = -\int_{\partial X_t} \rho_t(s) u_j^b(s)^2 ds.$$

Let $A(t)$ be the area of S_t. By Weyl's law, $E_j(t) \sim c\frac{j}{A(t)}$. Since the area of X_t grows linearly, we have on average $\dot{E}_j \sim -C\frac{E_j}{A(t)}$. Theorem 6.2 gives the asymptotics individually for almost all eigenvalues. Let

$$f_j(t) = \int_{\partial X_t} \rho_t(s) |u_j^b(t;s)|^2 ds.$$

Then $\dot{E}_j = -E_j f_j$. Then Theorem 6.2 implies that $|u_j^b|^2 \to \frac{1}{A(t)}$ weakly on the boundary along a subsequence of density one. QUE is the hypothesis that this occurs for the entire sequence, i.e.

$$f_j(t) \to \frac{k}{A(t)} > 0, \quad k := \int_{\partial S_t} \rho_t(s) ds.$$

Hence,

$$\frac{\dot{E}_j}{E} = -kA(t)(1 + o(1)), \quad j \to \infty.$$

Hence there is a lower bound to the velocity with which eigenvalues decrease as $A(t)$ increases. Eigenvalues can therefore not concentrate in the fixed quasi-mode intervals $[n^2 - K, n^2 + K]$ for all t. But then there are only a bounded number of eigenvalues in this interval; so Proposition 7.2 implies QUE for the other X_t. A more detailed analysis shows that QUE holds for almost all t.

8. Matrix elements of Fourier integral operators

Our main focus in this survey goal is on the limits of diagonal matrix elements (12) of pseudodifferential operators A or order zero relative to an orthonormal basis $\{\varphi_j\}$ of eigenfunctions. Difficult as it is, the limiting behavior of such matrix elements is one of the more accessible properties of eigenfunctions. To obtain more information, it would be useful to expand the class of operators A for which one can study matrix elements. In §5, we were essentially expanding the class from classical polyhomogeneous pseudodifferential operators to those in the "small-scale" calculus, i.e. supported in h^δ tubes around closed geodesics. Another way to expand the class of A is to consider matrix elements of Fourier integral operators. Motivation to consider matrix elements of Fourier integral operators comes partly from the fact that Hecke operators are Fourier integral operators, and this section is a preparation for the next section on Hecke operators §9. Other examples arise in quantum ergodic restriction theorems (see [**TZ3**]). Details on the claims to follow will be given in a forthcoming paper.

By an FIO (Fourier integral operator) we mean an operator whose Schwartz kernel of F may be locally represented as a finite sum of oscillatory integrals,

$$K_F(x,y) \sim \int_{\mathbb{R}^N} e^{i\varphi(x,y,\theta)} a(x,y,\theta) d\theta$$

for some homogeneous phase φ and amplitude a. The only example we have seen so far in this survey is the wave group U^t and of course pseudo-differential operators.

The phase is said to generate the canonical relation

$$C = \{(x, \varphi'_x, y, -\varphi'_y) : \varphi'_\xi(x,y,\theta) = 0\} \subset T^*M \times T^*M.$$

The oscillatory integral also determines the principal σ_F of F, a 1/2 density along C. The data (C, σ_F) determines F up to a compact operator. In the case of the wave group, C is the graph of g^t and the symbol is the canonical volume half-density on the graph. In the case of ψDO's, the canonical relation is the diagonal (i.e. the graph of the identity map) and the symbol is the one defined above. We denote by $I^0(M, \times M, C)$ the class of Fourier integral operators of order zero and canonical relation C.

Hecke operators are also FIO's of a simple kind. As discussed in the next section, in that case C is a local canonical graph, i.e. the projections

$$\pi_X : C \to T^*M, \quad \pi_Y : C \to T^*M$$

are branched covering maps. Equivalently, C is the graph of a finitely many to one correspondence $\chi : T^*M \to T^*M$. Simpler examples of the same type include (i) $F = T_g$ is translation by an isometry of a Riemannian manifold (M, g) possessing an isometry, or (ii) $Tf(x) = \sum_{j=1}^k f(g_j x) + f(g_j^{-1} x)$ on S^n corresponding to a finite set $\{g_1, \dots, g_k\}$ of isometries. The latter has been studied by Lubotzky-Phillips-Sarnak.

8.1. Matrix elements as linear functionals. The matrix elements

$$\rho_j(F) := \langle F\varphi_j, \varphi_j \rangle \tag{82}$$

define continuous linear functionals

$$\rho_j : I^0(M \times M, C) \to \mathbb{C} \tag{83}$$

with respect to the operator norm. It is simple to see that any limit ρ_∞ of the sequence of functionals ρ_j are functionals only of the principal symbol data (C, σ_F) which is bounded by the supremum of σ_F. Thus, as in the pseudo-differential case, ρ_∞ is a measure on the space $\Omega_C^{1/2}$ of continuous half-densities on C.

The limiting behavior of $\rho_j(F)$ depends greatly on whether $[F, \sqrt{\Delta}] = 0$ (or is of negative order) or whether it has the same order as F. In effect, this is the issue of whether the canonical relation C underlying F is invariant under the geodesic flow or not. For instance, if C is the graph of a canonical transformation χ and if χ moves the energy surface $S_g^* M = \{(x, \xi) : |\xi|_g = 1\}$ to a new surface $\chi(S_g^* M)$ disjoint from $S_g^* M$ then $\langle F\varphi_j, \varphi_j \rangle \to 0$. On the other hand, $\langle Ae^{it\sqrt{\Delta}}\varphi_j, \varphi_j \rangle = e^{it\lambda_j}\langle A\varphi_j, \varphi_j \rangle$ so one gets the same quantum limits for $F = Ae^{it\sqrt{\Delta}}$ as for A but must pick sparser subsequences to get the limit.

Hecke operators have the property that $[F, \Delta] = 0$. This implies that the canonical relation C is invariant under g^t. In general this is the only condition we need to study weak limits. The eigenfunction linear functionals ρ_k are invariant in the the sense that $\rho_k(U^{-t}FU^t) = \rho_k(F)$.

PROPOSITION 8.1. *Let $C \subset T^*M \times T^*M$ be a local canonical graph, equipped with the pull-back of the symplectic volume measure on T^*M. Assume C is invariant under g^t. Let $F \in I^0(M \times M, C)$ and identify its symbol with a scalar function relative to the graph half-density. Let ρ_∞ be a weak* limit of the functionals ρ_j on $I^0(M \times N, C)$. Then there exists a complex measure ν on $SC = C \cap S^*M \times S^*M$ of mass ≤ 1 such that*

$$\rho_\infty(F) = \int_{SC} \sigma_F d\nu,$$

which satisfies $g_*^t d\nu = d\nu$.

The proof is similar to that in the ψDO case.

We note that $I^0(M \times M, C)$ is a right and left module over $\Psi^0(M)$. Given $F \in I^0(M \times M, C)$ we consider the operators $AF, FA \in I^0(M \times M, C)$ where $A \in \Psi^0(X)$. We obtain a useful improvement on Proposition 8.1.

PROPOSITION 8.2. *With the same hypotheses as in Proposition 8.1. Assume further that F is self-adjoint and that $\rho_j(AF) \sim \rho_j(FA)$. Then we have:*

$$\rho_\infty(\pi_X^* a\sigma) = \rho_\infty(\pi_Y^* a\sigma).$$

This holds since

$$\rho_\infty(\pi_X^* a\sigma) \sim \rho_j(AF) = \langle AF\varphi_j, \varphi_j \rangle = \overline{\langle F^* A^* \varphi_j, \varphi_j \rangle} = \overline{\langle FA^* \varphi_j, \varphi_j \rangle} \sim \pi_Y^* a\sigma$$

since $\sigma_{A^*} \sim \overline{\sigma_A}$ and since $\overline{\sigma_F} = \sigma_F$.

Let us consider some simple examples. First, suppose that $F = T_g + T_g^*$ where g is an isometry. Then C is the union of the graph of the lift of g to T^*M (by its derivative) and the inverse graph. The above Proposition then says that $g_*\nu = \nu$. This is obvious by Egorov's theorem, and we can regard the Proposition as a generalization of Egorov's theorem to symplectic correspondences. As a second example, let $F = U^t$. Then $\langle AF\varphi_j, \varphi_j \rangle = e^{it\lambda_j}\langle A\varphi_j, \varphi_j \rangle$, so we see that sequences with unique quantum limits are sparser than in the pseudo-differential case.

9. QUE of Hecke eigenfunctions

Our next topic is Hecke eigenfunctions. A detailed treatment would have to involve adelic dynamics, higher rank measure rigidity in the presence of positive entropy [**LIND, LIND2**] and L-functions [**HS, Sound1**]. Both areas are outside the scope of this survey. Fortunately, Lindenstrauss has written an expository article on adelic dynamics and higher rank rigidity for an earlier Current Developments in Mathematics [**LIND2**], and Soundararajan has recently lectured on his L-function results in a 2009 Clay lecture. Sarnak has recently written exposition of the new results of Soundararajan and Holowinsky-Soundararajan results is [**Sar3**].

So we continue in the same vein of explaining the microlocal (phase space) features of the Hecke eigenvalue problem. The distinguishing feature of Hecke eigenfunctions is that they have a special type of symmetry. As a result, their quantum limit measures have a special type of symmetry additional to geodesic flow invariance. The analysis of these additional symmetries is a key input into Lindenstrauss' QUE result. The main purpose of this section is to derive the exact symmetry. In fact, it appears that the exact symmetry was not previously determined; only a quasi-invariance condition of [**RS**] has been employed.

We begin by recalling the definition of a Hecke operator. Let Γ be a co-compact or cofinite discrete subgroup of $G = PSL(2, \mathbb{R})$ and let $\mathbf{X}_\Gamma$ be the corresponding compact (or finite area) hyperbolic surface. An element $g \in G, g \notin \Gamma$ is said to be in the commensurator $Comm(\Gamma)$ if

$$\Gamma'(g) := \Gamma \cap g^{-1}\Gamma g$$

is of finite index in Γ and $g^{-1}\Gamma g$. More precisely,

$$\Gamma = \bigcup_{j=1}^{d} \Gamma'(g)\gamma_j, \quad \text{(disjoint)},$$

or equivalently

$$\Gamma g \Gamma = \bigcup_{j=1}^{d} \Gamma \alpha_j, \quad \text{where } \alpha_j = g\gamma_j.$$

It is also possible to choose α_j so that $\Gamma g \Gamma = \bigcup_{j=1}^d \alpha_j \Gamma$. We then have a diagram of finite (non-Galois) covers:

$$\begin{array}{ccccc} & & \Gamma'(g)\backslash \mathbf{H} & & \\ & \pi \swarrow & & \searrow \rho & \\ \Gamma\backslash \mathbf{H} & & \Longleftrightarrow & & g^{-1}\Gamma g\backslash \mathbf{H}. \end{array} \tag{84}$$

Here,

$$\pi(\Gamma'(g)z) = \Gamma z, \quad \rho(\Gamma'(g)z) = g\Gamma g^{-1}(g\gamma_j z) = g\Gamma z,$$

where in the definition of ρ any of the γ_j could be used. The horizontal map is $z \to g^{-1}z$.

A Hecke operator

$$T_g f(x) = \sum_{j=1}^{d} f(g\gamma_j x) : L^2(\mathbf{X}_\Gamma) \to L^2(\mathbf{X}_\Gamma) \tag{85}$$

is the Radon transform $\rho_* \pi^*$ of the diagram. We note as a result that

$$T_{g^{-1}} u(z) = u(g^{-1}z)$$

takes Γ-invariant functions to $g^{-1}\Gamma g$-invariant functions. The Hecke operator is thus an averaging operator over orbits of the Hecke correspondence, which is the multi-valued holomorphic map

$$C_g(z) = \{\alpha_1 z, \dots, \alpha_d z\}.$$

Its graph

$$\Gamma_g = \{(z, \alpha_j z) : z \in \mathbf{X}_\Gamma)\}$$

is an algebraic curve in $\mathbf{X}_\Gamma \times g^{-1}Xg$.

The Hecke operators T_g form a commutative ring as g varies over $Comm(\Gamma)$. By a Hecke eigenfunction is meant a joint eigenfunction of Δ and the ring of Hecke operators:

$$T_g u_j = \mu_j(g) u_j, \quad \Delta u_j = -\lambda_j^2 u_j.$$

9.1. Quantum limits on $\Gamma'(g)\backslash\mathbf{H}$. We now considered symmetries of quantum limits of Hecke eigenfunctions. We assume u_j is a Hecke eigenfunction. It is convenient to modify the definitions of the functionals ρ_k as follows:

$$\begin{cases} \rho'_j(A) = \frac{\langle A\pi^* u_j, \rho^* u_j\rangle}{\langle \pi^* u_j, \rho^* u_j\rangle} = \dfrac{\langle \rho_* A \pi^* u_j, u_j\rangle}{\langle \pi^* u_j, \rho^* u_j\rangle}, \\ \sigma'_j(A) = \dfrac{\langle A\pi^* u_j, \pi^* u_j\rangle}{\langle \pi^* u_j, \rho^* u_j\rangle}, \\ \tau'_j(A) = \dfrac{\langle A\pi^* u_j, \rho^* u_j\rangle}{\langle \rho^* u_j, \rho^* u_j\rangle} \end{cases} \tag{86}$$

on $A \in \Psi^0(\Gamma'(g)\backslash\mathbf{H})$. Thus, if $A_0 \in \Psi^0(\mathbf{X}_\Gamma)$ we have,

$$\frac{\langle A_0 \rho_* \pi^* u_j, u_j\rangle}{\langle \rho_* \pi^* u_j, u_j\rangle} = \langle A_0 u_j, u_j\rangle. \tag{87}$$

Since Hecke operators are FIO's, limit measures of $\langle AT_g u_j, u_j\rangle$ and $\langle T_g A u_j, u_j\rangle$ live on the canonical relation underlying T_g, which is the lift of the graph of the Hecke correspondence to $T^*\mathbf{X}_\Gamma$. It is convenient to restate the result by forming a second diagram

$$
\begin{array}{ccccc}
 & & \Gamma_g \subset \mathbf{X}_\Gamma \times g^{-1}\Gamma g\backslash \mathbf{H} & & \\
 & \pi_1 \swarrow & & \searrow \rho_1 & \\
\Gamma\backslash \mathbf{H} & & \Longleftrightarrow & & g^{-1}\Gamma g\backslash \mathbf{H}.
\end{array}
\tag{88}
$$

Here, π_1, ρ_1 are the natural projections. In the case of arithmetic groups such as $SL(2,\mathbb{Z})$, $Comm(\Gamma)$ is dense in G, i.e. there are many Hecke operators. The map

$$\iota : \Gamma'(g)\backslash \mathbf{H} \to \Gamma_g$$

defined by

$$\iota(z) = (\pi(z), \rho(z)),$$

is a local diffeomorphic parametrization, and $\pi_1 \iota = \pi, \rho_1 \iota = \rho$.

The general Proposition 8.2 then specializes to:

PROPOSITION 9.1. *If ω' is a weak* limit of ρ'_j, and if ω is the weak* limit on $\mathbf{X}_\Gamma$ then $\pi_* \omega' = \rho_* \omega' = \omega$.*

We may reformulate the proposition as follows: if we lift measures and functions to the universal cover $\mathbb{H}$, then the data of the Hecke limit measures are a finite set of real signed measures $\{d\nu_k\}$ on the different copies of $S^*\mathbb{H}$. The latter must project to the former under both projections, so we get the exact invariance property,

$$\nu = \sum_{j=1}^{d} \nu_{\alpha_j} = \sum_{j=1}^{d} \alpha_{j*} \nu_{\alpha_j}, \tag{89}$$

for some complex measures ν_j on the j sheet of the cover. The sum of the masses of the ν_j must add up to one. This is simpler to see for Hecke operators on S^2. There we also have some number r of copies of S^2 and r isometries g_j and define the Hecke operator (of Lubotzky-Phillips-Sarnak) as in (85). Then a quantum limit measure ν_0 on S^*S^2 is the projection of r limit measures on the copies of S^*S^2 under both projections.

Previously, only a quasi-invariance property of quantum limit measures of Hecke eigenfunctions was proved by Rudnick-Sarnak [**RS**], and used in [**BL, W3**] and elsewhere. In the notation above, it may be stated as

$$\nu(E) \leq d \sum_j |\alpha_{j*} \nu_{\alpha_j}(E)|. \tag{90}$$

As a simple application of (89) (which was also clear from the quasi-invariance condition), we sketch a proof that a quantum limit measure ν of a sequence of Hecke eigenfunctions for $\Gamma = SL(2,\mathbb{Z})$ cannot be a periodic orbit measure μ_γ. If it were, it would have to be $\sum \nu_j$ as above. Each ν_j is an invariant signed measure so by ergodic decomposition, each must have the form $c_j\mu_\gamma + \tau_j$ where $\sum c_j = 1$, where τ_j are singular with respect to μ_γ and $\sum_j \tau_j = 0$. But also the images of ν_j under the α_j must have the same property. This is only possible if $\alpha_j(\gamma) = \gamma$ up to a translation by $\beta \in \Gamma$ whenever $\nu_j \neq 0$. At least one of the $c_j \neq 0$. At this point, one must actually consider the elements α_j in the Hecke operator. In the case of $PSL(2,\mathbb{Z})$ they are parabolic elements plus one elliptic element. But only a hyperbolic element can fix a geodesic (its axis).

It would be interesting to see if one can determine how the ν_{α_j} are related to each other. Lindenstrauss's QUE result (in the next section) implies that that $(T_g)_*\nu = \nu$, which appears to say that the ν_{α_j} are all the same. An apriori proof of this would simplify the proof of QUE.

9.2. QUE results of Lindenstrauss, Soundararajan (and Holowinsky-Soundararajan). We now briefly recall the QUE results on Hecke eigenfunctions.

THEOREM 9.2. *(E. Lindenstrauss* [**LIND**]*) Let* $\mathbf{X}_\Gamma$ *be a compact arithmetic hyperbolic surface. Then QUE holds for Hecke eigenfunctions, i.e. the entire sequence of Wigner distributions of Hecke eigenfunctions tends to Liouville measure.*

In the non-compact finite area case of $\Gamma = PSL(2,\mathbb{Z})$, Lindenstrauss proved:

THEOREM 9.3. *(E. Lindenstrauss* [**LIND**]*) Let* $\mathbf{X}_\Gamma$ *be the arithmetic hyperbolic surface defined by a congruence subgroup. Then any weak* limit of the sequence of Wigner distributions of Hecke eigenfunctions is a constant multiple of Liouville measure (the constant may depend on the sequence).*

Note that the quantum ergodicity theorem does not apply in this finite area case. In the arithmetic case there is a discrete spectrum corresponding to cuspidal eigenfunctions and a continuous spectrum corresponding to Eisenstein series. The following was proved in [**Z8**]:

THEOREM 9.4. *Let* $\mathbf{X}_\Gamma$ *be the arithmetic hyperbolic surface defined by a congruence subgroup, and let* $\{\varphi_j\}$ *be any orthonormal basis of cuspidal eigenfunctions. Then for any pseudo-differential operator of order zero and with compactly supported symbol,*

$$\frac{1}{N_c(\lambda)} \sum_{j:\lambda_j \leq \lambda} |\langle A\varphi_j, \varphi_j\rangle - \omega(A)|^2 \to 0,$$

where $N_c(\lambda) \sim |\mathbf{X}_\Gamma|\lambda^2$ *is the number of cuspidal eigenvalues* $\leq \lambda$.

Recent work of Soundararajan that the constant equals one, i.e. there is no mass leakage at infinity. Hence the theorem is improved to

THEOREM 9.5. *(E. Lindenstrauss* [**LIND**], *Soundarajan* [**Sound1**]*) Let* $\mathbf{X}_\Gamma$ *be the arithmetic hyperbolic surface defined by a congruence subgroup.*

Then any weak limit of the sequence of Wigner distributions of Hecke eigenfunctions is is normalized Liouville measure.*

Holowinsky [**Hol**] and Holowinsky-Soundararajan [**HS**] have proven QUE results for holomorphic forms (i.e. for the discrete series; see §1.15.4). However, the methods are entirely different; they are based on the theory of L-functions and Poincaré series. The Hecke property is exploited through multiplicativity of Fourier coefficients.

9.3. Hecke QUE and multiplicities. A natural question (raised at the Clay talk of Soundararajan) is whether QUE holds for *any* orthonormal basis of eigenfunctions of an arithmetic quotient if it holds for the Hecke eigenbasis and if the discrete spectrum of Δ has bounded multiplicities. The proof that this is true is a simple modification of Proposition 7.2 and we pause to sketch it now.

PROPOSITION 9.6. *Suppose that one orthonormal basis of Δ is QUE, and that the eigenvalues have uniformly bounded multiplicities. Then all orthonormal bases are QUE.*

PROOF. As in Proposition 7.2, consider sequences $\{(\lambda_{i_r}, \lambda_{j_r}),\ i_r \neq j_r\}$ of pairs of eigenvalues of $\sqrt{\Delta}$ with $\lambda_{i_r} = \lambda_{j_r}$. We then have that $dW_{i_r, j_r} \to 0$.

This has implications for all orthonormal basis as long as eigenvalue multiplicities are uniformly bounded. We run the argument regarding modes and quasi-modes but use the QUE orthonormal basis in the role of modes and any other orthonormal basis $\{\psi_k\}$ in the role of quasi-modes.

LEMMA 9.7. *If the eigenvalue multiplicities are uniformly bounded and there exists a sequence of eigenfunctions $\{\psi_k\}$*

- *(i)* $n(k) \leq C,\ \forall\, k$;
- *(ii)* $\langle A\psi_k, \psi_k\rangle \to \int_{S^*M} \sigma_A d\mu$ *where* $d\mu \neq d\mu_L$.

Then no orthonormal basis of eigenfunctions of Δ is QUE.

PROOF. We argue by contradiction. If $\{\varphi_j\}$ were QUE, we would have (in the notation of (77):

$$\langle A\psi_k, \psi_k\rangle = \sum_{i,j=1}^{n(k)} c_{kj} c_{ki} \langle A\varphi_i, \varphi_j\rangle + o(1)$$

$$= \sum_{j=1}^{n(k)} c_{kj}^2 \langle A\varphi_j, \varphi_j\rangle + \sum_{i \neq j=1}^{n(k)} c_{kj} c_{ki} \langle A\varphi_i, \varphi_j\rangle + o(1)$$

$$= \int_{S^*M} \sigma_A d\mu_L + o(1),$$

by Proposition 9.6. This contradicts (ii). In the last line we used $\sum_{j=1}^{n(k)} |c_{kj}|^2 = 1 + o(1)$, since $||\psi_k||_{L^2} = 1$. □

10. Variance estimates: Rate of quantum ergodicity and mixing

A quantitative refinement of quantum ergodicity is to ask at what rate the sums in Theorem 4.1(1) tend to zero, i.e. to establish a rate of quantum ergodicity. More generally, we consider "variances" of matrix elements. For diagonal matrix elements, we define:

$$V_A(\lambda) := \frac{1}{N(\lambda)} \sum_{j:\lambda_j \leq \lambda} |\langle A\varphi_j, \varphi_j) - \omega(A)|^2. \tag{91}$$

In the off-diagonal case one may view $|\langle A\varphi_i, \varphi_j\rangle|^2$ as analogous to $|\langle A\varphi_j, \varphi_j) - \omega(A)|^2$. However, the sums in (57) are double sums while those of (91) are single. One may also average over the shorter intervals $[\lambda, \lambda+1]$.

10.1. Quantum chaos conjectures. It is implicitly conjectured by Feingold-Peres in [**FP**] (11) that

$$|\langle A\varphi_i, \varphi_j\rangle|^2 \simeq \frac{C_A(\frac{E_i - E_j}{\hbar})}{2\pi\rho(E)}, \tag{92}$$

where $C_A(\tau) = \int_{-\infty}^{\infty} e^{-i\tau t}\langle V^t\sigma_A, \sigma_A\rangle dt$. In our notation, $\lambda_j = \hbar^{-1}E_j$ (with $\hbar = \lambda_j^{-1}$ and $E_j = \lambda_j^2$) and $\rho(E)dE \sim dN(\lambda)$.

On the basis of the analogy between $|\langle A\varphi_i, \varphi_j\rangle|^2$ and $|\langle A\varphi_j, \varphi_j) - \omega(A)|^2$, it is conjectured in [**FP**] that

$$V_A(\lambda) \sim \frac{C_{A-\omega(A)I}(0)}{\lambda^{n-1}vol(\Omega)}.$$

The idea is that $\varphi_\pm = \frac{1}{\sqrt{2}}(\varphi_i \pm \varphi_j)$ have the same matrix element asymptotics as eigenfunctions when $\lambda_i - \lambda_j$ is sufficiently small. But then $2\langle A\varphi_+, \varphi_-\rangle = \langle A\varphi_i, \varphi_i\rangle - \langle A\varphi_j, \varphi_j\rangle$ when $A^* = A$. Since we are taking a difference, we may replace each matrix element by $\langle A\varphi_i, \varphi_i\rangle$ by $\langle A\varphi_i, \varphi_i\rangle - \omega(A)$ (and also for φ_j). The conjecture then assumes that $\langle A\varphi_i, \varphi_i\rangle - \omega(A)$ has the same order of magnitude as $\langle A\varphi_i, \varphi_i\rangle - \langle A\varphi_j, \varphi_j\rangle$.

10.2. Rigorous results for Hecke eigenfunctions. Let $\mathbf{X}_\Gamma$ be the arithmetic modular surface, with $\Gamma = SL(2, \mathbb{Z})$, and let T_n, $n \geqslant 1$ denote the family of Hecke operators. In addition to the Laplace eigenvalue problem (28), the Maass-Hecke eigenforms are eigenfunctions of the Hecke operators,

$$T_n\varphi = \lambda_\varphi(n)\varphi.$$

In the cusp, they have Fourier series expansions

$$\varphi(z) = \sqrt{y}\sum_{n \neq 0} \rho_\varphi(n) K_{ir}(2\pi|n|y)e(nx)$$

where K_{ir} is the K-Bessel function and $\rho_\varphi(n) = \lambda_\varphi(n)\rho_\varphi(1)$. A special case of the variance sums (91) is the configuration space variance sum,

$$S_\psi(\lambda) = \sum_{\lambda_j \leq \lambda} |\int \psi\varphi^2 dV|^2$$

Following the notation of [**LS2**] and [**Zh**],we put $d\mu_n = \varphi_{r_n}^2 dV$.

Let $\psi \in C^\infty_{0,0}(S^*\mathbf{X}_\Gamma)$ and consider the distribution of

$$\frac{1}{\sqrt{T}} \int_0^T \psi(g^t(x,\xi))dt.$$

Ratner's central limit theorem for geodesic flows implies that this random variable tends to a Gaussian with mean 0 and the variance given by the non-negative Hermitian form on $C^\infty_{0,0}(Y)$:

$$V_{\mathrm{C}}(\varphi,\psi) = \int_{-\infty}^{\infty} \int_{\Gamma\backslash SL(2,\mathbf{R})} \varphi\left(g\begin{pmatrix} e^{\frac{t}{2}} & 0 \\ 0 & e^{-\frac{t}{2}} \end{pmatrix}\right) \overline{\psi(g)} dg dt.$$

Restrict V_C to $C^\infty_{0,0}(X)$. Different irreducible representations are orthogonal under the quadratic form. Maass cusp forms φ are eigenvectors of V_C. If the Laplace eigenvalue is $\frac{1}{4} + r^2$, then the V_C eigenvalue is

$$V_{\mathrm{C}}(\varphi,\varphi) = \frac{|\Gamma(\frac{1}{4} - \frac{ir}{2})|^4}{2\pi|\Gamma(\frac{1}{2} - ir)|^2}$$

In [**LS, LS2**], Luo and Sarnak studied the quantum variance for holomorphic Hecke eigenforms, i.e. holomorphic cusp forms in $S_k(\Gamma)$ of even integral weight k for Γ. In this setting, the weight of the cusp form plays the role of the Laplace eigenvalue. They proved that for $\varphi, \psi \in C^\infty_{0,0}(X)$, as the weight $K \to \infty$,

$$\sum_{k \leq K} \sum_{f \in H_k} L(1, \mathrm{Sym}^2 f)\mu_f(\varphi)\overline{\mu_f(\psi)} \sim B(\varphi,\psi)K,$$

where $B(\varphi,\psi)$ is a non-negative Hermitian form on $C^\infty_{0,0}(X)$. B is diagonalized by the orthonormal basis of Maass-Hecke cusp forms and the eigenvalues of B at ψ_j is $\frac{\pi}{2}L(\frac{1}{2},\psi_j)$. For the notation $L(1,\mathrm{Sym}^2)$ we refer to [**LS2, Zh**].

In [**Zh**], P. Zhao generalized the result in [**LS, LS2**] to Maass cusp forms. Let $\varphi_j(z)$ be the j-th Maass-Hecke eigenform, with the Laplacian eigenvalues $\frac{1}{4} + r_j^2$.

THEOREM 10.1. [**Zh**] *Fix any $\epsilon > 0$. Then we have*

$$\sum_{t_j} (e^{-(\frac{r_j - T}{T^{1-\epsilon}})^2} + e^{-(\frac{r_j + T}{T^{1-\epsilon}})^2}) L(1, \mathrm{sym}^2 \varphi_j)\mu_j(\varphi)\overline{\mu_j}(\psi)$$
$$= T^{1-\epsilon}V(\varphi,\psi) + O(T^{\frac{1}{2}+\epsilon}),$$

V is diagonalized by the orthonormal basis $\{\varphi_j\}$ of Maass-Hecke cusp forms and the eigenvalue of V at a Maass-Hecke cusp form φ is

$$L(\frac{1}{2}, \varphi) V_{\mathbb{C}}(\varphi, \varphi).$$

It would be desirable to remove the weights. The unweighted version should be (P. Zhao, personal communication; to appear)

$$\sum_{\lambda_j \leq \lambda} \mu_j(\varphi) \overline{\mu}_j(\psi) \sim \lambda V(\varphi, \psi). \tag{93}$$

10.3. General case. The only rigorous result to date which is valid on general Riemannian manifolds with hyperbolic geodesic flow is the logarithmic decay:

THEOREM 10.2. [**Z4**] *For any (M, g) with hyperbolic geodesic flow,*

$$\frac{1}{N(\lambda)} \sum_{\lambda_j \leq \lambda} |(A\varphi_j, \varphi_j) - \omega(A)|^{2p} = O(\frac{1}{(\log \lambda)^p}).$$

It is again based on Ratner's central limit theorem for the geodesic flow. The logarithm reflects the exponential blow up in time of remainder estimates for traces involving the wave group associated to hyperbolic flows and thus the necessity of keeping the time less than the Ehrenfest time (18). It would be surprising if the logarithmic decay is sharp for Laplacians. However, R. Schubert shows in [**Schu, Schu2**] that the estimate is sharp in the case of two-dimensional hyperbolic quantum cat maps. Hence the estimate cannot be improved by semi-classical arguments that hold in both settings.

10.4. Variance and Patterson-Sullivan distributions. When $\int_{S^*\mathbf{X}_\Gamma} a d\mu_L = 0$, the Luo-Sarnak variance results for Hecke eigenfunctions have the form,

$$\frac{1}{N(\lambda)} \sum_{j:|r_j| \leq \lambda} \left|\langle a, dW_{ir_j} \rangle\right|^2 \sim \frac{1}{\lambda} V(a, a). \tag{94}$$

Even if one knew that asymptotics of this kind should occur, it does not seem apriori obvious that the bilinear form on the right side should be invariant under g^t because $V(a, a)$ occurs in the $\frac{1}{\lambda}$ term in the variance asymptotics and only only knows that the Wigner distributions are invariant modulo terms of order $\frac{1}{\lambda_j}$. A possible way to see that the bilinear form should be g^t-invariant is via the Patterson-Sullivan distributions. From (51), we apparently have

$$\frac{1}{N(\lambda)} \sum_{j:|r_j| \leq \lambda} |\langle a, dW_{ir_j} \rangle|^2 = \frac{1}{N(\lambda)} \sum_{j:|r_j| \leq \lambda} |\langle a, d\hat{PS}_{ir_j} \rangle|^2 + O(\lambda^{-2}). \tag{95}$$

Indeed, the difference is of the form $\frac{1}{N(\lambda)}\sum_{j:|r_j|\leq\lambda}\frac{2}{r_j}\text{Re}\,\langle a, d\hat{PS}_{ir_j}\rangle\cdot$ $\langle L_2 a, d\hat{PS}_{ir_j}\rangle + O(\lambda^{-2})$. The sum in the last expression also seems to be of order $O(\lambda^{-2})$ (e.g. one can go back and express each factor of PS with one of W_{ir_j} and apply the Luo-Sarnak asymptotics again). Since $\hat{PS}_{ir_j}$ is g^t-invariant, it follows that $V(a,a)$ must be g^t invariant. This argument is equally valid on any compact hyperbolic surface, but of course there is no proof of such asymptotics except in the arithmetic Hecke case. It would be interesting to draw further relations between $V(a,a)$ and PS_{ir_j}.

11. Entropy of quantum limits on manifolds with Anosov geodesic flow

So far, the results on quantum limits have basically used the symmetry (or invariance) properties of the limits. But generic chaotic systems have no symmetries. What is there to constrain the huge number of possible limits?

The recent results of Anantharaman [**A**], Anantharaman-Nonnenmacher [**AN**] and Rivière [**Riv**], give a very interesting answer to this question (see also [**ANK**]) for (M,g) with Anosov geodesic flow. They use the quantum mechanics to prove lower bounds on entropies of quantum limit measures. The lower bounds eliminate many of the possible limits, e.g. they disqualify finite sums of periodic orbit measures.

The purpose of this section is to present the results of [**A**] and [**AN**] in some detail. Both articles are based on a hyperbolic dispersive estimate (called the *main estimate* in [**A**]) which, roughly speaking, measures the norm of quantized cylinder set operators in terms of $\hbar$ and the length of the cylinder. But they use the estimate in quite different ways. In [**A**], it is used in combination with an analysis of certain special covers of S^*M by cylinder sets that are adapted to the eigenfunctions. One of the important results is an estimate on the topological entropy $h_{top}(\text{supp}(\nu_0))$ of the support of any quantum limit measure. In [**AN**] the key tool is the "entropic uncertainty principle". It leads to a lower bound for the Kolomogorv-Sinai entropy $h_{KS}(\nu_0)$ of the quantum limit.

There now exist several excellent and authoritative expositions of the KS entropy bounds and the entropic uncertainty principle [**A2, ANK, AN2, CV3**] in addition to the well written initial articles [**A, AN**] (which take considerable care to give intuitive explanations of technical steps). For this reason, we emphasize the approach in [**A**]. We closely follow the original references in discussing heuristic reasons for the lower bounds and outlining rigorous proofs. We also discuss earlier entropy lower bounds of Bourgain-Lindenstrauss [**BL**] and Wolpert [**W3**] in the case of Hecke eigenfunctions.

Before stating the results, we review the various notions of entropy.

11.1. KS entropy. We first recall the definition of the KS entropy of an invariant probability measure μ for the geodesic flow. Roughly speaking, the entropy measures the average complexity of μ-typical orbits. In the

Kolmogorov-Sinai entropy, one starts with a partition $\mathcal{P} = (E_1, \ldots, E_k)$ of $S_g^* M$ and defines the Shannon entropy of the partition by $h_{\mathcal{P}}(\mu) = \sum_{j=1}^k \mu(E_j) \log \mu(E_j)$. Under iterates of the time one map g of the geodesic flow, one refines the partition from the sets E_j to the cylinder sets of length n:

$$\mathcal{P}^n := \{\mathcal{P}^n_\alpha := E_{\alpha_0} \cap g^{-1} E_{\alpha_1} \cap \cdots \cap g^{-n+1} E_{\alpha_{n-1}} : \ [\alpha_0, \ldots, \alpha_{n-1}] \in \mathbf{N}^n\}.$$

One defines $h_n(\mathcal{P}, \mu)$ to be the Shannon entropy of this partition and then defines $h_{KS}(\mu, \mathcal{P}) = \lim_{n \to \infty} \frac{1}{n} h_n(\mu, \mathcal{P})$. Then $h_{KS}(\mu) = \sup_{\mathcal{P}} h_{KS}(\mu, \mathcal{P})$.

The measure $\mu(E_{\alpha_0} \cap g^{-1} E_{\alpha_1} \cap \cdots \cap g^{-n+1} E_{\alpha_{n-1}})$ is the probability with respect to μ that an orbit successively visits $E_{\alpha_0}, E_{\alpha_1}, \ldots$. The entropy measures the exponential decay rate of these probabilities for large times.

11.2. Symbolic coding and cylinder sets. In the (M, g) setting, we fix a partition $\{M_k\}$ of M and a corresponding partition $T^* M_k$ of $T^* M$. Let P_{α_k} be the characteristic function of $T^* M_k$. (Later it must to be smoothed out). Let $\Sigma = \{1, \ldots, \ell\}^{\mathbb{Z}}$ where ℓ is the number of elements of the partition $\mathcal{P}^{0)}$. To each tangent vector $v \in S^* M$, one can associated a unique element $I(v) = (\alpha_k) \in \Sigma$ so that $g^n v \in P_{\alpha_j}$ for all $n \in \mathbb{Z}$. This gives a symbolic coding map $I : S^* M \to \Sigma$. The time one map g^1 then conjugates under the coding map to the shift $\sigma((\alpha_j)) = ((\alpha_{j+1}))$ on admissible sequences, i.e. sequences in the image of the coding map. An invariant measure ν_0 thus corresponds to a shift invariant measure μ_0 on Σ.

A cylinder set $[\alpha_0, \ldots, \alpha_{n-1}] \subset \Sigma$ of length n is the subset of Σ formed by sequences with the given initial segment. The set of such cylinder sets of length n is denoted Σ_n. Cylinder sets for Σ are not the same as cylinder sets for g^1, which have the form $P_{\alpha_0} \cap g^{-1} P_{\alpha_1} \cap \cdots \cap g^{-n} P_{\alpha_n}$.

The μ_0 measure of a Σ-cylinder set is by definition,

$$\mu_0([\alpha_0, \ldots, \alpha_n]) = \nu_0(P_{\alpha_0} \cap g^{-1} P_{\alpha_1} \cap \cdots \cap g^{-n+1} P_{\alpha_{n-1}}).$$

11.3. Bowen balls. Cylinder sets are closely related to Bowen balls, i.e. balls in the Bowen metric. For any smooth map $f : X \to X$, the Bowen metric at time n is defined by

$$d_n^f(x, y) = \max_{0 \leq i \leq n-1} d(f^i x, f^i y).$$

Let $B_n^f(x, r)$ be the r ball around x in this metric.

In the continuous time case of the geodesic flow g^t, one defines the d_T metric

$$d_T(\rho, \rho') = \max_{t \in [0T]} d(g^t(\rho), g^t(\rho')). \tag{96}$$

The Bowen ball $B_T(\rho, r)$ at time T is the r- ball in this metric.

As with cylinder sets, if ρ, ρ' lie in $B_T(\rho, \epsilon)$ then their orbits are ϵ-close for the interval $[0, T]$. This is not quite the same as running through the same elements of a partition but for large T the balls and cylinders are rather similar. This is because the geodesic flow g^t stretches everything by a factor of e^t in the unstable direction, and contracts everything by e^{-t} in the stable direction; it preserves distances along the geodesic flow. In variable curvature and in higher dimensions, the "cube" becomes a rectangular parallelopiped whose axes are determined by the Lyapunov exponents.

In the case of geodesic flows of compact hyperbolic manifolds of constant curvature -1, Bowen balls $B_T(\rho, r)$ are roughly of radius re^{-T} in the unstable direction, and r in the stable direction and geodesic flow directions. In the hyperbolic case of G/Γ with co-compact Γ, the tube (in the notation of §1.15.1) is

$$B_T(\rho, \epsilon)) = a((-r, r))n_-(r, r))n_+((-e^{-T}r, e^{-T}r)). \tag{97}$$

To make the ball symmetric with respect to the stable and unstable directions is to make it symmetric with respect to time reversal. One uses the time interval $[-T/2, T/2]$ instead of $[0, T]$ and defines the new distance,

$$d'_T(\rho, \rho') = \max_{t \in -[T/2, T/2]} d'(g^t(\rho), g^t(\rho')), \tag{98}$$

We then denote the r ball by $B'_T(\rho, r)$. In the hyperbolic case,

$$B'_T(\rho, \epsilon)) = a((-\epsilon, \epsilon))n_-((-e^{-T/2}\epsilon, e^{-T/2}\epsilon))n_+((-e^{-T/2}\epsilon, e^{-T/2}\epsilon)). \tag{99}$$

Thus, the Bowen ball in the symmetric case (in constant curvature) is a ball (or cube) of radius re^{-T} in the transverse direction to the geodesic flow. The length along the flow is not important.

11.4. Brin-Katok local entropy. Kolmogorov-Sinai entropy of an invariant measure is related to the local dimension of the measure on Bowen balls. This is the approach in Bourgain-Lindenstrauss [**BL**] and in [**LIND**], and stems from work of Young and Ledrappier-Young.

Define the local entropy on an invariant measure μ for a map f by

$$h_\mu(f, x) = \lim_{\delta \to 0} \limsup_{n \to \infty} \frac{-\log \mu(B_n^f(x, \delta))}{n} = \lim_{\delta \to 0} \liminf_{n \to \infty} \frac{-\log \mu(B_n^f(x, \delta))}{n}.$$

Brin-Katok proved that both limits exist, that the local entropy $h_\mu(f, x)$ is f invariant and

$$h_\mu(f) = \int_X h_\mu(f, x) d\mu(x).$$

The definition and result is similar in the geodesic flow case. The local entropy of μ at ρ is defined by,

$$\begin{aligned} &\lim_{\epsilon \to 0} \liminf_{T \to \infty} -\frac{1}{T} \log \mu(B_T(\rho, \epsilon)) \\ &\quad = \lim_{\epsilon \to 0} \limsup_{T \to \infty} -\frac{1}{T} \log \mu(B_T(\rho, \epsilon)) := h_{KS}(\mu, \rho). \end{aligned}$$

11.5. Ergodic decomposition. Let ν be an invariant measure. It may be expressed as a convex combination of ergodic measures, which are extreme points of the compact convex set of invariant measures. There is a concrete formula,

$$\nu = \int \nu_x^{\mathcal{E}} d\nu(x) \tag{100}$$

for its ergodic decomposition, where $\nu_x^{\mathcal{E}}$ is the orbital average through $x \in G/\Gamma$. It is a fact that $\nu_x^{\mathcal{E}}$ ergodic for ν-a.e. x. Then the local entropy is

$$h_{a(t)}(\tilde{\mu}_x^{\mathcal{E}}) = \lim_{\epsilon \to 0} \frac{\tilde{\mu}(T(x,\epsilon))}{\log \epsilon}, \quad a.s.\tilde{\mu}. \tag{101}$$

The Brin-Katok theorem states that the global KS entropy is

$$h(\nu) = \int h(\nu_x^{\mathcal{E}}) d\nu(x).$$

11.6. Topological entropy of an invariant subset. Anantharaman's first entropy bound refers to the topological entropy of an invariant set F. Then, by definition,

$$\begin{aligned} h_{top}(F) \leq \lambda \iff & \forall \delta > 0,\ \exists C > 0 : F \text{ can be covered by at most} \\ & Ce^{n(\lambda+\delta)} \text{ cylinders of length } n, \ \ \forall n. \end{aligned} \tag{102}$$

11.7. Bourgain-Lindenstrauss entropy bound. Before discussing the work of Anantharman-Nonnenmacher, we review a simpler entropy bound of Bourgain-Lindenstrauss. After initial work of Wolpert [**W3**], Bourgain-Lindenstrauss [**BL**] obtained a strong lower bound on entropies of quantum limit measures associated to Hecke eigenfunctions. It was an ingredient in Lindenstrauss' QUE theorem [**LIND**]. It is simple to state and illustrates the use of the local entropy and local dimensions to measure the KS entropy.

The setting is a compact or finite area arithmetic hyperbolic quotient. We have not reviewed local entropy in the non-compact finite area case, but proceed by analogy with the compact case. Then

$$B(\epsilon, \tau_0) = a((-\tau_0, \tau_0))n_-(\epsilon, \epsilon))n_+((-\epsilon, \epsilon))$$

is a Bowen ball around the identity element and $xB(\epsilon, \tau_0)$ is its left translate by $x \in G$. Here $\epsilon = re^{-T/2}$.

THEOREM 11.1. [**BL**] *Let $\Gamma \subset SL(2, \mathbb{Z})$ be a congruence subgroup. Let μ be a quantum limit. Then for any compact subset $K \subset \Gamma \backslash G$ and any $x \in K$,*

$$\mu(xB(\epsilon, \tau_0)) \leq \epsilon^{2/9}.$$

In the flow-box notation above,

$$\mu_n(B_T(\rho,\epsilon)) \le Ce^{-T/9}. \tag{103}$$

This implies that any ergodic component of any quantum limit has entropy $\ge \frac{1}{9}$.

COROLLARY 11.2. [**BL**] *Almost every component of a quantum limit measure has entropy $\ge 1/9$. The Hausdorff dimenson of the support of the limit measure is $\ge 1 + 1/9$.*

Theorem 11.1 is a consequence of the following uniform upper bound for masses in small tubes around geodesic segments in configuration space:

THEOREM 11.3. [**BL**] *Let $\Gamma \subset SL(2,\mathbb{Z})$ be a congruence subgroup. Let μ be a quantum limit. Then for any compact subset $K \subset \Gamma\backslash G$ and any $x \in K$, and for any Hecke eigenfunction φ_{λ_j},*

$$\int_{xB(r,\tau_0)} |\varphi_{\lambda_j}|^2 dV \le r^{2/9}.$$

One may rewrite this result in terms of a Riesz-energy of the quantum limit measures:

COROLLARY 11.4. *Let μ be a quantum limit of a sequence of Hecke-Maass eigenfunctions. Then for $\kappa < 2/9$,*

$$\int_M \int_M \frac{d\mu(x)d\mu(y)}{d_M(x,y)^{\kappa+1}} < \infty.$$

We briefly sketch the idea of the proof since it makes an interesting contrast to the Anantharaman-Nonnenmacher bound. They prove that there exists a set W of integers of size $\epsilon^{-2/9}$ so that for $n \in W$ one has (roughly) that the translates $yB(\epsilon,\tau_0)$ of the small balls by y in the Hecke correspondence for T_n are all pairwise disjoint. Since

$$|\varphi_\lambda(y)|^2 \le \sum_{z \in T_n(y)} |\varphi_\lambda(z)|^2,$$

one has

$$\int_{xB(\epsilon,\tau_0)} |\varphi_\lambda(y)|^2 dV \le C \int_{xB(\epsilon,\tau_0)} \sum_{z\in T_n(y)} |\varphi_\lambda(z)|^2 dV$$

$$= \sum_{z\in T_n(x)} \int_{zB(\epsilon,\tau_0)} |\varphi_\lambda(z)|^2 dV.$$

Now sum over $n \in W$ and use the disjointness of the small balls $zB(\epsilon, \tau_0)$ and the L^2 normalization of the eigenfunction to obtain

$$\int_{xB(\epsilon,\tau_0)} |\varphi_\lambda(y)|^2 dV \leq \frac{1}{|W|} \sum_{n \in W} \sum_{z \in T_n(x)} \int_{zB(\epsilon,\tau_0)} |\varphi_\lambda(z)|^2 dV$$

$$\leq C\epsilon^{2/9} \int_X |\varphi_\lambda^2 dV \leq C\epsilon^{2/9}.$$

11.8. Anantharman and Anantharaman-Nonnenacher lower bound. We now turn to the lower bounds on entropy in the general Anosov case given in [**A**] and in [**AN, ANK**]. The mechanisms proving the lower bounds and the results are somewhat different between the two articles. The result of [**A**] gives a lower bound for the topological entropy $h_{top}(supp \nu_0)$ of the support of a quantum limit while the main result of [**AN**] gives a lower bound on the metric or Kolmogorov Sinai entropy $h_g = h_{KS}$ of the quantum limits. For the definition of $h_{top}(F)$, see (102).

THEOREM 11.5. [**A**] *Let* $\nu_0 = \int_{S^*M} \nu_0^x d\nu_0(x)$ *be the ergodic decomposition of* ν_0. *Then there exists* $\kappa > 0$ *and two continuous decreasing functions* $\tau : [0,1] \to [0,1]$ *and* $\vartheta : (0,1] \to \mathbb{R}_+$ *with* $\tau(0) = 1, \vartheta(0) = \infty$, *such that*

$$\nu_0[\left\{x : h_g(\nu_0^x) \geq \frac{\Lambda}{2}(1-\delta)\right\}] \geq (\frac{\kappa}{\vartheta(\delta)})^2 (1 - \tau(\delta)).$$

Hence

- $h_g(\nu_0) > 0$;
- $h_{top}(supp\ \nu_0) \geq \frac{\Lambda}{2}$.

It follows that a positive proportion of the ergodic components of ν_0 must have KS entropy arbitrarily close to $\frac{\Lambda}{2}$.

This is proved as a consequence of a Proposition which may have other applications:

THEOREM 11.6. [**A**] *(Proposition 2.0.4) With the same notation as in Theorem 11.5 and Remark 11.10. Let* F *be a subset with* $h_{top}(F) \leq \frac{\Lambda}{2}(1-\delta)$. *Then,*

$$\nu_0(F) \leq \left(1 - (\frac{\kappa}{\vartheta(\delta)})^2\right)(1 - \tau(\delta)) < 1.$$

Hence supp ν_0 *cannot equal* F.

In [**A**] Theorem 1.1.2, there is a generalization to quasi-modes which introduces a constant c. For simplicity we only consider eigenfunctions.

In the subsequent article of Anantharaman-Nonnenmacher, [**AN, ANK**] the authors obtain a quantitative lower bound on the KS entropy:

THEOREM 11.7. [**AN, ANK**] *Let μ be a semiclassical measure associated to the eigenfunctions of the Laplacian on M. Then its metric entropy satisfies*

$$h_{KS}(\mu) \geq \left| \int_{S^*M} \log J^u(\rho) d\mu(\rho) \right| - \frac{(d-1)}{2} \lambda_{\max}, \tag{104}$$

where $d = \dim M$. $\lambda_{\max} = \lim_{|t| \to \infty} \frac{1}{t} \log \sup_{\rho \in \mathcal{E}} |dg^t_\rho|$ is the maximal expansion rate of the geodesic flow on $\mathcal{E}$ (5) and J^u is the unstable Jacobian determinant (3).

When M has constant sectional curvature -1, the theorem implies that

$$h_{KS}(\mu) \geq \frac{d-1}{2}. \tag{105}$$

They state the conjecture

CONJECTURE 11.8. [**AN**] *Let (M, g) have Anosov geodesic flow. Then for any quantum limit measure ν_0,*

$$h_g(\nu_0) \geq \frac{1}{2} \left| \int_{S^*M} \log J^u(v) d\nu_0(v) \right|.$$

This conjecture has recently been proved by G. Rivière for surfaces of negative curvature [**Riv**].

This conjecture does not imply QUE. In fact the counterexamples in [**FNB, FN**] to QUE in the case of "quantum cat maps" satisfy the condition of the conjecture. So there is some chance that the conjecture is best possible for (M, g) with ergodic geodesic flow.

To the author's knowledge, the corollary that quantum limits in the Anosov case cannot be pure periodic orbit measures has not been proved a simpler way that by applying the theorems above. In §11.21 we will go over the proof in that special case and see how it ties together with mass concentration of eigenfunctions around hyperbolic closed geodesics in §5.

11.9. Problem with semi-classical estimates and Bowen balls. As a first attempt to estimate KS entropies of quantum limits, one might try to study the local entropy formula as in Bourgain-Lindenstrauss. Let $B_T(\rho, \epsilon)$ be the small tube around $\rho \in S^*M$, i.e. of length ϵ in the flow direction and $e^{-T/2}$ in the stable/unstable directions. As in the Bourgain-Lindenstrauss bound, one would like to understand the decay of $\mu_n(B_T(\rho, \epsilon))$ as $T \to \infty$ and $n \to \infty$. In the brief expository article [**AN2**], the authors give the heuristic estimate

$$\mu_n(B_T(\rho, \epsilon)) \leq C \lambda_n^{\frac{d-1}{2}} e^{-\frac{(d-1)T}{2}}, \quad d = \dim M \tag{106}$$

on the local dimensions of the quantum measures for a sequence of eigenfunctions. Unlike the uniform Bourgain-Lindenstrauss estimate, the estimate is λ_n-dependent and is only non-trivial for times $T > \log \lambda_n$. But then the ball radius (in the stable/unstable directions) is $e^{-T} = \lambda_n^{-1/2}$, which is just below the minimal scale allowed by the uncertainty principle (§1.4). To obtain a useful entropy bound one needs to take $T = 2 \log \lambda_n$ and then $e^{-T} = \lambda_n^{-1}$, well below the Heisenberg uncertainty threshold. So although it is attractive, the local entropy is difficult to use. Recall that Bourgain-Lindenstrauss used the many elements of the Hecke correspondences to move the small ball around so that it almost fills out the manifold, and then used the L^2 normalization to estimate the total mass. There doesn't seem to exist a similar mechanism to build up a big mass from the small balls in the general Anosov case.

11.10. Some important parameters. Two important parameters κ, ϑ appear here and in the statements and proofs in [**A**]. They represent time scales relative to the Ehrenfest time:

- κ: Semi-classical estimates only work when $n \leq \kappa|\log \hbar|$, i.e. $\kappa|\log \hbar|$ plays the role of the Ehrenfest time.
- ϑ: The main estimate is only useful when $h^{-d/2} e^{-n\Lambda/2} << 1$ or $n \geq \vartheta|\log \hbar|$.

11.11. Cylinder set operators in quantum mechanics. To get out of the too-small Bowen ball impasse, Anantharaman [**A**] and Ananthraman-Nonnenmacher study "quantum measures" of *cylinder sets* $\mathcal{C}$ rather than small balls. The emphasis (and results) on quantum cylinder sets is one of the key innovations in [**A**]. The classical cylinder sets are not directly involved in the main estimates of [**A, AN**]. Rather they are quantized as cylinder set operators. Then their "quantum measure" or quantum entropy is studied. To distinguish notationally between classical and quantum objects, we put a $\hat{\cdot}$ on the quantum operator.

To define quantum cylinder set operators, one quantizes the partition to define a smooth quantum partition of unity $\hat{P}_k$ by smoothing out the characteristic functions of M_k. Let $\alpha = [\alpha_0, \ldots, \alpha_{n-1} \in \Sigma_n$ be a cylinder set in sequence space Σ. The corresponding quantum cylinder operator is

$$\hat{P}_\alpha = \hat{P}_{\alpha_{n-1}}(n-1)\hat{P}_{\alpha_n-2}(n-2)\cdots\hat{P}_{\alpha_0}, \tag{107}$$

where $\alpha = [\alpha_0, \ldots, \alpha_{n-1}]$ where

$$\hat{P}(k) = U^{*k}\hat{P}U^k. \tag{108}$$

Here, $U = e^{i\sqrt{\Delta}}$ is the propagator at unit time (or in the semi-classical framework, $U = e^{i\hbar\Delta/2}$ (see [**A**] (1.3.4)).

The analogue of the measure of a cylinder set in quantum mechanics is given by the matrix element of the cylinder set operator in the energy state, transported to Σ. We use the semi-classical notation of [**A**], but it is easily converted to homogeneous notation.

DEFINITION: (See [**A**], (1.3.4)) Let $\psi_\hbar$ be an eigenfunction of Δ. Define the associated quantum "measure" of cylinder sets $\mathcal{C} = [\alpha_0, \dots, \alpha_{n-1}] \in \Sigma_n$ by

$$\mu_\hbar([\alpha_0, \dots, \alpha_n]) = \langle \hat{P}_{\alpha_n}(n) \cdots \hat{P}_{\alpha_0}(0)\psi_\hbar, \psi_\hbar\rangle. \tag{109}$$

Thus, one "quantizes" the cylinder set $\mathcal{C} = [\alpha_0, \dots, \alpha_{n-1}]$ as the operator

$$\hat{\mathcal{C}} = U^{-(n-1)}\hat{P}_{\alpha_{n-1}}U\hat{P}_{\alpha_{n-2}} \cdots U\hat{P}_{\alpha_0} \tag{110}$$

For each eigenfunction, one obtains a linear functional

$$\mu_\hbar([\alpha_0, \dots, \alpha_{n-1}]) = \rho_\hbar(\hat{\mathcal{C}}) \tag{111}$$

in the notation of states of §3. The "quantum measures" are shift invariant, i.e.

$$\mu_\hbar([\alpha_0, \dots, \alpha_{n-1}]) = \mu_\hbar(\sigma^{-1}[\alpha_0, \dots, \alpha_{n-1}]).$$

This says,

$$\begin{aligned} &\langle U^{-(n-2)}P_{\alpha_{n-1}}U^{n-2} \cdots UP_{\alpha_0}U^{-1}\varphi_\hbar, \varphi_\hbar\rangle \\ &= \langle U^{-(n-1)}P_{\alpha_{n-1}}U^{n-1} \cdots UP_{\alpha_0}\varphi_\hbar, \varphi_\hbar\rangle, \end{aligned}$$

which is true since $\rho_h(A) = \langle A\varphi_\hbar, \varphi_\hbar\rangle$ is an invariant state.

The matrix element $\langle \hat{P}_\alpha\psi_\hbar, \psi_\hbar\rangle$ is the probability (amplitude) that the particle in state $\psi_\hbar$ visits the phase space regions $P_{\alpha_0}, P_{\alpha_1}, \dots, P_{\alpha_{n-1}}$ at times $0, 1, \dots, n-1$.

Thus, the states $\rho_\hbar$ or alternatively the Wigner distributions of $\psi_\hbar$ are transported to Σ to define linear functionals $\mu_\hbar$ on the span of the cylinder functions on Σ. They *exactly* invariant under the shift map.

The functionals $\mu_\hbar$ are not positive measures since $\hat{\mathcal{C}}_\hbar$ is not a positive operator. However as in the proof of the quantum ergodicity Theorem 4.1, one has the Schwartz inequality,

$$|\rho_\hbar(\hat{\mathcal{C}})|^2 \leq \rho_\hbar(\hat{\mathcal{C}}^*\hat{\mathcal{C}}), \tag{112}$$

reflecting the positivity of the state $\rho_\hbar$.

11.12. Main estimate for cylinder sets of Σ. The so-called *main estimate* of [**A**] implies that for every cylinder set $\mathcal{C} \in \Sigma_n$, one has

$$|\mu_\hbar(\mathcal{C})| \leq C_\beta h^{-d/2}e^{-n\Lambda/2}(1+O(\epsilon))^n, \quad \text{uniformly for } n \leq \beta\,|\log\hbar|. \tag{113}$$

Here, β can be taken arbitrarily large. This estimate is similar in spirit to (106) but circumvents the uncertainty principle. It is one of the essential quantum mechanical (or semi-classical) ingredient for getting a lower bound on $h_{top}(supp\nu_0)$ of supports of quantum limit measures (see §11.15).

11.13. Quantization of cylinder sets versus quantized cylinder operators. Let us explain how (113) gets around the uncertainty principle. A fundamental feature of quantization is that it is not a homomorphism. Thus, the quantization of the (smoothed) characteristic function of the cylinder set $P_{\alpha_0} \cap g^{-1} P_{\alpha_1} \cap \cdots \cap g^{-n} P_{\alpha_n}$ is very different from $\hat{P}_{\alpha_{n-1}} U \hat{P}_{\alpha_{n-2}} \cdots U \hat{P}_{\alpha_0}$, the ordering which occurs in the quantum cylinder operator $\hat{P}_\alpha$ in (107). The problematic Bowen ball estimate (106) would arise if one quantized the Bowen ball (or cylinder set) in the first sense and then took its matrix element. Quantizing in the other order makes the main estimate possible.

11.14. Quantized cylinder operators and covering numbers. As motivation to study quantum measures of cylinder sets, Anantharaman observes that the main estimate suggests an obstruction to a quantum measure concentrating on a set F with $h_{top}(F) < \frac{\Lambda}{2}$. For each n such a set can be covered by $C e^{n(\frac{\Lambda}{2}(1-\delta)}$ cylinder sets of length n. By the main estimate (113), each of the cylinder sets has quantum measure $|\mu_\hbar(\mathcal{C})| \leq C e^{-d/2 \log \hbar - n\Lambda/2}$ uniformly for $n \leq \kappa \, |\log \hbar|$ for any κ. As yet, the combination does not disprove that supp$\nu_0 = F$, i.e. $\nu_0(F) = 1$, since the large factor of $h^{-d/2}$ is not cancelled in the resulting estimate $\mu_\hbar(F) \leq C e^{n(\frac{\Lambda}{2}(1-\delta) - d/2 \log \hbar - n\Lambda/2}$. An additional idea is needed to decouple the large time parameter n into independent time parameters. Roughly speaking, this is done by a sub-multiplicative property. In §11.20, the ideas in the above paragraph are made more precise and effective.

11.15. Upper semi-continuity, sub-multiplicativity and sub-additivity. Before getting into the details of (the sketch of) the proof, it is useful to consider why there should exist a non-trivial lower bound on $h_{KS}(\nu_0)$ for a quantum limit measure? What constrains a quantum limit measure to have higher entropy than some given invariant measure?

First, the classical entropy of an invariant is upper semi-continuous with respect to weak convergence. This is in the right direction for proving lower entropy bounds for quantum limits. If one can obtain a lower bound on the "entropies" of the quantum measures μ_n, then the entropy of the limit can only jump up. Of course, the quantum measures are not g^t invariant measures and so upper semi-continuity does not literally apply. One needs to define a useful notion of entropy for the quantum measures and prove some version of upper semi-continuity for it. The USC property of the classical entropy suggests that there should exist some analogue on the quantum level.

In [**A**], entropies are studied via special covers by cylinder sets. For these, there is a sub-multiplicative estimate on the number of elements in the cover.

In [**AN**], a quite different idea is used: the authors employ a notion of quantum entropy due to Maassen-Uffink and prove a certain lower bound for the quantum entropy called the entropy uncertainty inequality. It is the origin of the lower bound for the quantum entropies and the KS entropy in Theorem 11.7 (see §11.17). As a replacement for the USC property of the entropy, a certain sub-additivity property for the quantum entropy is proved (see §11.18).

In the next three subsections, we go over the main ingredients in the proof. Then we give a quick sketch of the full proof.

11.16. Main estimate in more detail. The main estimate in [**A**] (Theorem 1.3.3) was refined in [**AN, ANK**]. To state the results, we need some further notation. Let $n_E(\hbar)$ denote the Ehrenfest time

$$n_E(\hbar) = \left[\frac{(1-\delta)|\log \hbar|}{\lambda_{\max}}\right]$$

where the brackets denote the integer part and δ is a certain small number arising in energy localization (for the definition of $\lambda_{\max}$ see (5)). The authors also introduce a discrete "coarse-grained" unstable Jacobian

$$J_1^u(\alpha_0, \alpha_1) := \sup\{J^u(\rho) : \rho \in T^*\Omega_{\alpha_0} \cap \mathcal{E}^\epsilon : \; g^t\rho \in T^*\Omega_{\alpha_1}\},$$

for $\alpha_0, \alpha_1 = 1, \ldots, K$. Here, Ω_j are small open neighborhoods of the partition sets M_j. For a sequence $\alpha = (\alpha_0, \ldots, \alpha_{n-1})$ of symbols of length n, one defines

$$J_n^u(\alpha) := J_1^u(\alpha_0, \alpha_1) \cdots J_1^u(\alpha_{n-2}, \alpha_{n-1}).$$

THEOREM 11.9. *(See Theorem 3.5 of* [**ANK**]*) Given a partition $\mathcal{P}^{(0)}$ and $\delta, \delta' > 0$ small enough, there exists $\hbar_{\mathcal{P}^{(0)},\delta,\delta'}$ such that, for any $\hbar \leq \hbar_{\mathcal{P}^{(0)},\delta,\delta'}$, for any positive integer $n \leq n_E(\hbar)$, and any pair of sequences α, α' of length n,*

$$\left\| \hat{P}_{\alpha'}^* \, U^n \, \hat{P}_\alpha \, \mathrm{Op}(\chi^{(n)}) \right\| \leq C \, \hbar^{-(d-1+c\delta)} \sqrt{J_n^u(\alpha) J_n^u(\alpha')} \,. \tag{114}$$

Here, $d = \dim M$ and the constants c, C only depend on (M, g).

There is a more refined version using a sharper energy cutoff in [**AN**].

THEOREM 11.10. *Given a partition $\mathcal{P}^{(0)}$, $\kappa > 0$ and $\delta > 0$ small enough, there exists $\hbar_{\mathcal{P}^{(0)},\delta,\kappa}$ such that, for any $\hbar \leq \hbar_{\mathcal{P}^{(0)},\delta,\kappa}$, for any positive integer $n \leq \kappa |\log \hbar|$, and any pair of sequences α, α' of length n,*

$$\left\| \hat{P}_{\alpha'}^* \, U^n \, \hat{P}_\alpha \, \mathrm{Op}(\chi^{(n)}) \right\| \leq C \, \hbar^{-(\frac{d-1}{2}+\delta)} \, e^{-(d-1)n} (1 + O(\hbar^\delta))^n \,. \tag{115}$$

Here, $d = \dim M$.

To prove this, one shows that any state of the form $\mathrm{Op}(\chi^{(*)})\Psi$ can be decomposed as a superposition of essentially $\hbar^{-\frac{(d-1)}{2}}$ normalized Lagrangian states, supported on Lagrangian manifolds transverse to the stable leaves of the flow. The action of the operator $\hat{P}_\alpha$ on such Lagrangian states is intuitively as follows: each application of U stretches the Lagrangian in the unstable direction (the rate of elongation being described by the unstable Jacobian) whereas each multiplication by $\hat{P}_{\alpha_j}$ cuts off the small piece of the Lagrangian in the αth cell. This iteration of stretching and cutting accounts for the exponential decay. A somewhat more detailed exposition is in §11.22.

This estimate can be reformulated as a statement about matrix elements of the operators $\hat{P}_\alpha$ relative to eigenfunctions (see Theorem 3.1 of [**ANK**])

THEOREM 11.11. *Let $\hat{P}_k = \mathbf{1}^{sm}_{M_k}$. For any $\kappa > 0$ there exists $\hbar_\kappa > 0$ so that, uniformly for $\hbar < \hbar_\kappa$ and all $n \leq \kappa \, |\log \hbar|$, and for all $\alpha = (\alpha_0, \ldots, \alpha_{n-1}) \in [1, \kappa]^n \cap \mathbb{Z}^n$,*

$$||\hat{P}_\alpha \psi_\hbar|| \leq 2(2\pi\hbar)^{-d/2} e^{-n\frac{\Lambda}{2}} (1 + O(\epsilon))^n.$$

The cutoff operator $Op(\chi^{(n)})$ is supported near the energy surface λ_n^2, and the eigenfunctions are supported on their energy surfaces. Here, Λ is the "smallest" expansion rate (see (4)).

We recall that $\langle \hat{P}_\alpha \psi_\hbar, \psi_\hbar \rangle$ is the probability (amplitude) that the particle in state $\psi_\hbar$ visits the phase space regions $P_{\alpha_0}, P_{\alpha_1}, \ldots, P_{\alpha_{n-1}}$ at times $0, 1, \ldots, n-1$. The main estimates show that this probability decays exponentially fast with n, at rate $\frac{\Lambda}{2}$. However, the decay only starts near the Eherenfest time $n_1 = \frac{d|\log \hbar|}{\Lambda}$.

11.17. Entropy uncertainty principle. The source of the lower bound for $h_{KS}(\nu_0)$ is most simply explained in [**AN**]. It is based on the "entropic uncertainty principle" of Maassen- Uffink. There are several notions of quantum or non-commutative entropy, but for applications to eigenfunctions it is important to find one with good semi-classical properties.

Let $(\mathcal{H}, \langle .,. \rangle)$ be a complex Hilbert space, and let $||\psi|| = \sqrt{\langle \psi, \psi \rangle}$ denote the associated norm. The quantum notion of partition is a family $\pi = (\pi_k)_{k=1,\ldots,\mathcal{N}}$ of operators on $\mathcal{H}$ such that $\sum_{k=1}^{\mathcal{H}} \pi_k \pi_k^* = Id$. If $||\psi|| = 1$, the entropy of ψ with respect to the partition π is define by

$$h_\pi(\psi) = -\sum_{k=1}^{\mathcal{N}} ||\pi_k^* \psi||^2 \log ||\pi_k^* \psi||^2 .$$

We note that the quantum analogue of an invariant probability measure μ is an invariant state ρ, and the direct analogue of the entropy of the partition would be $\sum \rho(\pi_k \pi_k^*) \log \rho(\pi_k \pi_k^*)$. If the state is $\rho(A) = \langle A\psi, \psi \rangle$ then $\rho(\pi_k \pi_k^*) = ||\pi_k^* \psi||^2$.

The dynamics is generated by a unitary operator $\mathcal{U}$ on $\mathcal{H}$. We now state a simple version of the entropy uncertainty inequality of Maasen-Uffink. A more elaborate version in [**A, AN, ANK**] gives a lower bound for a certain "pressure".

THEOREM 11.12. *For any $\epsilon \geq 0$, for any normalized $\psi \in \mathcal{H}$,*

$$h_\pi(\mathcal{U}\psi) + h_\pi(\psi) \geq -2 \log c(\mathcal{U}) ,$$

where

$$c(\mathcal{U}) = \sup_{j,k} |\langle e_k, \mathcal{U} e_j \rangle|$$

is the supremum of all matrix elements in the orthonormal basis $\{e_j\}$*. In particular,* $h_\pi(\psi) \geq -\log c(\mathcal{U})$ *if* ψ *is an eigenfunction of* $\mathcal{U}$*.*

The intuitive idea is that if a unitary matrix has small entries, then each of its eigenvectors must have large Shannon entropy.

This uncertainty principle is applied to the entropies of the partitions defined above. For the propagator, we put $\mathcal{U} = e^{iT_E\sqrt{\Delta}}$ is the wave operator at the "Ehrenfest time" $T_E = \frac{\log \lambda}{\lambda_{\max}}$. Or in the semi-classical framework (where $h = \frac{1}{\lambda}$), one can use the Hamiltonian is $H = h^2\Delta$ and the time evolution $\mathcal{U} = e^{in_E(h)h\Delta}$ with $n_E(h) = \frac{\log \frac{1}{h}}{\lambda_{\max}}$.

Applied to an eigenfunction of $\mathcal{U}$, one has

$$h_\pi(\psi_h) = \sum_{|\alpha|=n_E} ||\hat{P}_\alpha^* \psi_h||^2 \log ||\hat{P}_\alpha^* \psi_h||^2, \tag{116}$$

and obviously

$$h_\pi(\mathcal{U}\psi) + h_\pi(\psi) = 2h_\pi(\psi).$$

On the right side,

$$c(\mathcal{U}) = \max_{|\alpha|=|\alpha'|=n_E} ||\hat{P}_{\alpha'} U^{n_E} \hat{P}_\alpha Op(\chi^{(n_E)})|| \tag{117}$$

where $\chi^{(n_E)}$ is a very sharp energy cutoff supported in a tubular neighborhood $\mathcal{E}^\epsilon := H^{-1}(1-\epsilon, 1+\epsilon)$ of $\mathcal{E} = S^*M$ of width $2h^{1-\delta}e^{n\delta}$ for a given $\delta > 0$.

11.18. Sub-additivity. Another key component in the proof is that the quantum entropy is almost sub-additive for $\hbar \leq |\log t|$.

Sub-additivity of the classical KS entropy follows from the concavity of the function $-x\log x$. It states that the sequence $\{h_n(\mu, P)\}$ is sub-additive, i.e.

$$h_{n+m}(\mu, P) \leq h_n(\mu, P) + h_m(\mu, P).$$

However, this is false for the quantum measures.

The correct statement is as follows: There exists a function $R(n_0, \hbar)$ such that $\lim_{\hbar\to 0} R(n_0, \hbar)$ for all $n_0 \in \mathbb{Z}$ and such that for all $n_0, n \in \mathbf{N}$ with $n_0 + n = \frac{(1-\delta)|\log \hbar|}{\lambda_{\max}}$, and for any normalized eigenfunction φ_λ, one has

$$h_{\mathcal{P}^{n_0+n}}(\varphi_\lambda) \leq h_{\mathcal{P}^{n_0}}(\varphi_\lambda) + h_{\mathcal{P}^n}(\varphi_\lambda) + R(n_0, \hbar). \tag{118}$$

11.19. Outline of proof of Theorem 11.7. We outline the proof from [**AN**]. Let

$$h_{\mathcal{P}^{(n)}}(\psi) = -\sum_{|\alpha|=n} ||\hat{P}_\alpha^* \psi||^2 \log ||\hat{P}_\alpha^* \psi||^2, \tag{119}$$

and let

$$h_{\mathcal{T}^{(n)}}(\psi) = -\sum_{|\alpha|=n} ||\hat{P}_\alpha \psi||^2 \log ||\hat{P}_\alpha \psi||^2, \tag{120}$$

(1) Let $K = \frac{1-\delta}{\lambda_{\max}}$. By the main estimate and the entropy uncertainty inequality, one has

$$h_{\mathcal{T}^{(n)}}(\psi_h) + h_{\mathcal{P}^{(n)}}(\psi_h) \geq 2(d-1)n + \frac{(d-1+2\delta)\lambda_{\max}}{(1-\delta)} n_E(\hbar) + O(1).$$

(2) Fix $n_0, n_E(\hbar)$. Write $n = qn_0 + r, r \leq n_0$. Then by sub-additivity,

$$\frac{h_{\mathcal{P}^{(n)}}(\psi_h)}{n} \leq \frac{h_{\mathcal{P}^{(n_0)}}(\psi_h)}{n_0} + \frac{h_{\mathcal{P}^{(r)}}(\psi_h)}{n_E(\hbar)} + \frac{R(n_0, \hbar)}{n_0}.$$

(3) As $\hbar \to 0$, this gives

$$\frac{h_{\mathcal{P}^{(n_0)}}(\psi_h)}{n_0} + \frac{h_{\mathcal{T}^{(n_0)}}(\psi_h)}{n_0} \geq 2(d-1)n + \frac{(d-1+2\delta)\lambda_{\max}}{(1-\delta)} n + O(1) - \frac{R(n_0, \hbar)}{2n_0} + O_{n_0}(1/n).$$

(4) Now take the limit $\hbar \to 0$. The expressions $||\hat{P}_\alpha \varphi_\lambda||^2$ and the Shannon entropies tend to their classical values. Hence, $h_{\mathcal{P}^{(n_0)}}(\psi_h)$ tends to the KS entropy of the quantum limit. Thus, the lower bound implies

$$\frac{h_{n_0}(\mu)}{n_0} \geq d - 1 - \frac{(d-1+2\delta)\lambda_{\max}}{2(1-\delta)}.$$

Since δ, δ' are arbitrary one can set them equal to zero. Then let $n_0 \to \infty$ to obtain the KS entropy of the quantum limit.

11.20. Sketch of proof of Theorem 11.5 [A]. The proof of Theorem 11.5 is less structural than the proof of Theorem 11.7. There is no magic weapon like the entropic uncertainty principle. In its place there is the construction of special covers by cylinder sets and counting arguments for the number of elements in the cover. These covering arguments probably have further applications and (although quantum) seem more geometric than the argument based on the entropic uncertainty principle.

We mainly discuss the statement that $h_{top}(\text{supp } \nu_0) \geq \frac{\Lambda}{2}$. We recall (§11.6) that an invariant set F is with $h_{top}(F) \leq \frac{\Lambda}{2}(1-\delta)$ can be covered by at $Ce^{N(\frac{\Lambda}{2}(1-\delta)}$ cylinder sets of length N. We further recall that, by the main estimate (113), each of the cylinder sets has quantum measure $|\mu_\hbar(\mathcal{C})| \leq Ce^{-d/2 \log \hbar - N\Lambda/2}$ uniformly for $N \leq \vartheta \ |\log \hbar|$ for any ϑ. We want to combine these estimates to show that $\nu_0(F) < 1$ for any quantum limit ν_0. But we cannot simply multiply the measure estimate for $\mathcal{C}$ in the cover times the number of cylinder sets in the cover. The main estimate being fixed, the only things left to work with are the covers.

As will be seen below, one needs to introduce special covers adapted to the eigenfunctions. The covering number of such covers has a sub-multiplicative property. We motivate it by going through a heuristic proof from [**A**]. We also add a new step that helps explain why special covers are needed. See also §11.21 where the pitfalls are explored when $F = \gamma$ (a closed hyperbolic orbit).

11.20.1. *Covers and times, I.* Let $F \subset \Sigma$ be a shift-invariant subset with $h_{top}(F) \leq \frac{\Lambda}{2}(1-\delta)$. Let $W_n \subset \Sigma_n$ be a cover of minimal cardinality of F by n cylinder-sets. Now introduce a new time $N >> n$ and define

$$\Sigma_N(W_n, \tau) := \begin{cases} \{N - \text{cylinders } [\alpha_0, \dots, \alpha_{N-1}] \text{ such that} \\ \\ \dfrac{\#\{j \in [0, N-n] : [\alpha_j, \dots, \alpha_{j+n-1}] \in W_n\}}{N-n+1} \geq \tau. \end{cases} \tag{121}$$

They correspond to orbits that spend a proportion $\geq \tau$ of their time in W_n. In terms of the (ergodic) Liouville measure (transported to Σ), for almost every point of S^*M (or its image under I), the proportion of the total time its orbit spends in W_n is the relative measure of (the union of the cylinder sets in) W_n. If this measure is smaller than τ, then $\Sigma_N(W_n, \tau)$ is a rare event as $N \to \infty$. Its cylinders satisfy $\sigma^j(\mathcal{C}) \cap W_n = \mathcal{C}$ for a proportion τ of $0 \leq j \leq N$. If one imagines partitioning the elements of W_n into N cylinders, then only a (relatively) sparse subset will go into $\Sigma_N(W_n, \tau)$. The next Lemma gives an upper bound on that number:

LEMMA 11.13. *(see Lemma 2.3.1 of* [**A**]*) For all n_0 there exist $n \geq n_0$ and N_0 so that for $N \geq N_0$ and $\tau \in [0, 1]$, the cover W_n of F of minimal cardinality satisfies,*

$$\#\Sigma_N(W_n, \tau) \leq e^{N\frac{3\Lambda\delta}{8}} e^{N h_{top}(F)} e^{(1-\tau)N(1+n)\log \ell}.$$

where ℓ is the number of elements in the partition.

If τ is chosen sufficiently close to 1, this implies:

COROLLARY 11.14. *(See (2.3.1) of* [**A**]*) If F is an invariant set with $h_{top}(F) < \frac{\Lambda}{2}(1-\delta)$, then*

$$\#\Sigma_N(W_n, \tau) \leq e^{N(\frac{\Lambda}{2}(1-\delta))} e^{(1-\tau)N(1+n)\log \ell}.$$

We now re-write this estimate in terms of the semi-classical parameter $\hbar$. If τ is sufficiently close to 1 and $\varepsilon > 0$ is sufficiently small, there exists $\vartheta > 0$ so that for $N \geq \vartheta |\log \hbar|$,

$$e^{N(\frac{\Lambda}{2}(1-\delta))} e^{(1-\tau)N(1+n)\log \ell} < (1-\varepsilon) h^{d/2} e^{N\Lambda/2}. \tag{122}$$

Indeed, let $\alpha = -\delta\frac{\Lambda}{2} + (1-\tau)(1+n)\log \ell$. For τ very close to 1 it is negative. We then choose ϑ so that $h^{-\alpha\vartheta} < (1-\varepsilon)h^{d/2}$, i.e. so that $|\alpha\vartheta| > d/2$.

By the main estimate (113) and (122), we obtain the following

COROLLARY 11.15. *(see* [**A**], *(2.0.1))*

For τ sufficiently close to 1 and ε close to zero, there exists $\vartheta > 0$ so that for $N \geq \vartheta |\log \hbar|$,

$$|\mu_\hbar(\Sigma_N(W_n, \tau))| \leq 1 - \varepsilon. \tag{123}$$

Here, $\mu_h(\Sigma_N(W_n,\tau))$ means $\mu_h(\bigcup_{\mathcal{C}\in\Sigma_N(W_n,\tau))} \mathcal{C})$.

11.20.2. *Heuristic Proof.* We now present a heuristic proof of Theorem 11.5 from Section 2 of [**A**] that explains the need for some rather technical machinery introduced below.

We recall that $\mu_\hbar$ is a shift (σ)-invariant measure on Σ. Let us temporarily pretend that $\mu_\hbar$ is also a positive measure. Then we would have for $N \geq \vartheta |\log \hbar|$,

$$\begin{aligned}
|\mu_\hbar(W_n)| &= |\frac{1}{N-n} \sum_{k=0}^{N-n-1} \mu_\hbar(\sigma^{-k} W_n)| \\
&= \left| \mu_\hbar \left(\frac{1}{N-n} \sum_{k=0}^{N-n-1} \mathbf{1}_{\sigma^{-k} W_n} \right) \right| \\
&\leq \sum_{\mathcal{C}\in\Sigma_N(W_n,\tau)} \mu_\hbar(\mathcal{C}) + \tau \sum_{\mathcal{C}\notin\Sigma_N(W_n,\tau)} \mu_\hbar(\mathcal{C}) \\
&= \mu_\hbar(\Sigma_N(W_n,\tau)) + \tau\mu_\hbar(\Sigma_N(W_n,\tau)^c) \\
&\leq (1-\tau)(1-\varepsilon) + \tau.
\end{aligned} \tag{124}$$

Here we used (123), and that the sum of the μ_h measures of $\Sigma_N(W_n, \tau)$ and its complement add to one. Since W_n is fixed, the weak convergence $\mu_\hbar \to \mu_0$ implies,

$$|\mu_0(W_n)| \;\; \leq \;\; (1-\tau)(1-\varepsilon)) + \tau < 1. \tag{125}$$

Since $F \subset W_n$, the same estimate applies to F. So F cannot be the support of μ_0. "QED"

In the above steps, we use that $\sigma^{-k} W_n \subset \Sigma_N(W_n, \tau)$ for $k \leq N - n - 1$ and that

$$\frac{1}{N-n} \sum_{k=0}^{N-n-1} \mathbf{1}_{\sigma^{-k} W_n} \leq 1, \quad \frac{1}{N-n} \sum_{k=0}^{N-n-1} \mathbf{1}_{\sigma^{-k} W_n} \leq \tau \text{ on } \Sigma_N(W_n,\tau)^c. \tag{126}$$

REMARK: There are two very innovative ideas in this argument. It is somewhat reminiscent of the averaging argument of Theorem 4.1, but there is no averaging over eigenvalues and no appeal to ergodicity (since there is no reason why a weak* limit of $\mu_\hbar$ should be an ergodic measure). The first idea is to study the special observables $\hat{P}_\alpha$. They are not expressed in the form $Op_\hbar(a)$, i.e. in terms of a symbol, and it would be counter-productive to do so. Thus, the argument of (135) takes place entirely on the quantum level and does not make use of any properties of the weak* limit (which are unknown); the semi-classical limit is taken after the inequality is proved. The key innovation is the third step, where the sets $\Sigma_N(W_n, \tau)$ are introduced and the sum is broken up into one over $\Sigma_N(W_n, \tau)$ and one over its complement. As is evident from (126), $\Sigma_N(W_n, \tau)^c$ arises naturally as an approximation to the τ-sublevel set of $\frac{1}{N-n}\sum_{k=0}^{N-n-1} \mathbf{1}_{\sigma^{-k}W_n}$. But note again that the the decomposition occurs in Σ_N and is used the on the quantum level. This is very different from using symbols of the $\hat{C}$ and their time averages to to define the corresponding subsets in S^*M!

The flaw in the argument is in the third line, where we pretended that $\mu_\hbar$ was a positive measure. We recall that $\mu_\hbar(\mathcal{C})$ converges to $\mu_0(\mathcal{C})$, which is a positive measure. But this is for a fixed $\mathcal{C}$. But it is also close to being a probability measure for cylinder sets of length $\leq \kappa|\log \hbar|$. But (123) is only valid for $N \geq \vartheta|\log \hbar|$. Since $\vartheta > \kappa$, one cannot use them both simultaneously (see §11.10 for κ, ϑ)

One can attempt to get around this problem using the positivity of the underlying state $\rho_\hbar$. Define the function

$$W_n^N := \frac{1}{N-n}\sum_{k=0}^{N-n-1} \mathbf{1}_{\sigma^{-k}W_n} : \Sigma \to \mathbb{R}_+.$$

Then in the third line we rigorously have,

$$|\mu_\hbar(W_n^N)| \leq \rho_\hbar\Big((\hat{W}_n^N)^*(\hat{W}_n^N)\Big)^{1/2} = ||\hat{W}_n^N Op_\hbar \psi_\hbar||. \tag{127}$$

Here, $\hat{W}_n^N$ is the quantization as a cylinder set operator. We further denote by $\widehat{\Sigma_N(W_n, \tau)}$ the operator equal to the sum of $\hat{\mathcal{C}}$ for $\mathcal{C} \in \Sigma_N(W_n, \tau)$. Then

$$\widehat{\Sigma_N(W_n, \tau)} + \widehat{\Sigma_N(W_n, \tau)^c} = I.$$

So (with $Op_\hbar(\chi)$ the energy cutoff),

$$\begin{aligned} ||\hat{W}_n^N Op_\hbar(\chi)\psi_\hbar|| &\leq ||\hat{W}_n^N \widehat{\Sigma_N(W_n, \tau)} Op_\hbar(\chi)\psi_\hbar|| \\ &\quad + ||\hat{W}_n^N \widehat{\Sigma_N(W_n, \tau)^c} Op_\hbar(\chi)\psi_\hbar||. \end{aligned} \tag{128}$$

By (126),

$$\begin{cases} ||\hat{W}_n^N \widehat{\Sigma_N(W_n, \tau)} Op_\hbar(\chi)\psi_\hbar|| \leq ||\widehat{\Sigma_N(W_n, \tau)} Op_\hbar(\chi)\psi_\hbar|| \\ \\ ||\hat{W}_n^N \widehat{\Sigma_N(W_n, \tau)}^c Op_\hbar(\chi)\psi_\hbar|| \leq \tau ||\widehat{\Sigma_N(W_n, \tau)}^c Op_\hbar(\chi)\psi_\hbar||. \end{cases}.$$

To complete the proof of the final equality, need an estimate of the form,

$$||\widehat{\Sigma_N(W_n, \tau)} Op_\hbar(\chi)\psi_\hbar|| \leq 1 - \theta. \tag{129}$$

We also need

$$||\widehat{\Sigma_N(W_n, \tau)}^c Op_\hbar \psi_\hbar|| \leq 1 - ||\widehat{\Sigma_N(W_n, \tau)} Op_\hbar \psi_\hbar|| + O(\hbar). \tag{130}$$

The latter holds formally since $\widehat{\Sigma_N(W_n, \tau)}, \widehat{\Sigma_N(W_n, \tau)}^c$ are semi-classically complementary projections. But the time $N \geq \vartheta |\log \hbar|$ may be too large for this semi-classical estimate.

The heuristic arguments are sufficiently convincing to motivate a difficult technical detour in which a certain sub-multiplicativity theorem is used to reduce the N in the argument from $N \geq \vartheta |\log \hbar|$ (above the Ehrenfest time) to $N \leq \kappa |\log \hbar|$ (below the Ehrenfest time). It is the analogue in [**A**] of the sub-additivity step in [**AN**], and its purpose is to allow the argument above to use only cylinder sets of length $\kappa |\log \hbar|$ (or balls of radius $>> \hbar^{1/2}$) where semi-classical analysis is possible. But one has to make sure that the covering number estimates in Lemma 11.13 and (122) do not break down. One has to choose special covers so that there are not too many elements in the cover of F. This step is another crucial innovation in [**A**], and in some ways the subtlest one.

11.20.3. $(\hbar, 1-\theta, n)$*-covers.* The sub-multiplicativity step requires a new concept and a new parameter θ. Fix $(\hbar, n, \theta)$ and consider a subset $W \subset \Sigma_n$.

DEFINITION: Say that W is an $(\hbar, 1-\theta, n)-$ cover of Σ if

$$|| \sum_{\mathcal{C} \in W^c} \hat{C}_\hbar Op_\hbar(\chi)\psi_\hbar|| \leq \theta. \tag{131}$$

Here, W^c is the complement of W; W is an $(\hbar, 1-\theta, n)$ cover of Σ if in this quantum sense the measure of W^c is < 1 (it is called "small" in [**A**] but in the end small means < 1; See the Remark at the end of §11.20.4). In other words, the union of the cylinder sets of W might not cover all of Σ (think S^*M), but the mass of the eigenfunctions $\psi_\hbar$ outside of the covered part is rather small. In this quantum sense, it is intuitively a union of cylinder sets whose measure is $\geq 1-\theta$. By definition, if W is an $(\hbar, 1-\theta, n)-$ cover, then it is an $(\hbar, 1-\theta', n)-$ cover for any $\theta' > \theta$.

The covering number of such a cover is defined by

$$N_\hbar(n, \theta) = \#\{W \colon W \text{ is a } (\hbar, 1-\theta, n) - \text{cover of } \Sigma\}. \tag{132}$$

The sub-multiplicative property is

LEMMA 11.16. *(sub-multiplicativity)*

$$N_\hbar(kn, k\theta(1+O(n\hbar^\alpha))) \leq N_\hbar(n,\theta)^k. \tag{133}$$

We will not attempt to explain this inequality but rather will point out its consequences. One is a lower bound on growth of counting numbers of minimal $(\hbar, 1-\theta, N)$-covers as N grows (Lemma 2.2.6 of [**A**]):

LEMMA 11.17. *Given $\delta > 0$, there exists ϑ so that:*

- *(i) For $N = \vartheta|\log\hbar|$, $N_\hbar(N,\theta) \geq (1-\theta)e^{N\frac{\Lambda}{2}(1-\frac{\delta}{16})}$;*
- *(ii) For $N = \kappa|\log\hbar|$, $\kappa \leq \vartheta$, $N_\hbar(N, \frac{\kappa}{\vartheta}\,\theta) \geq (1-\theta)^{\frac{\kappa}{\vartheta}} e^{N\frac{\Lambda}{2}(1-\frac{\delta}{16})}$;*

Estimate (ii) follows from (i) and sub-multiplicativity. The time in (ii) is now below the Ehrenfest time. The sub-multiplicative property then has the following consequence:

LEMMA 11.18. *(See [**A**], (2.3.2); compare (131)) Let κ, ϑ have the meanings in §11.10. Let $W_n \subset \Sigma_n$ be a cover of F by n-cylinders of minimal cardinality; and let $N = \kappa|\log\hbar|$. Then for any $\theta < 1$,*

$$\Big|\Big| \sum_{\mathcal{C}\in\Sigma_N(W_n,\tau)^c} \hat{C}_\hbar Op_\hbar(\chi)\psi_\hbar\Big|\Big| \geq \frac{\kappa}{\vartheta}\theta.$$

PROOF. (Sketch) The inequality goes in the opposite direction from that in the definition of $(\hbar, 1-\theta, N)$-covers in (11.19). Hence the Lemma is equivalent to saying that $\Sigma_N(W,\tau)$ is not a $(\hbar, 1-\frac{\kappa}{\vartheta}\theta, N)$-cover. This follows by comparing the upper bound in Lemma 11.13 with the lower bound in 11.17. If $\Sigma_N(W_n,\tau)$ were such a cover, one would have $\frac{\kappa}{\vartheta}|\log(1-\theta)| \geq N\frac{N\Lambda}{2}$, which can only occur for finitely many N if $\theta < 1$. The cardinality of $\Sigma_N(W,\tau)$ is too small to be $(\hbar, 1-\theta, N)$-cover when $N = \kappa|\log\hbar|$. □

The lower bound is a key step in the proof of Theorem 11.5. Intuitively, it gives a lower bound on eigenfunction mass outside the set of orbits which spend a proportion τ of their time in W_n. The lower bound on the mass outside $\Sigma_N(W_n,\tau)$ eventually prohibits the concentration of the quantum limit measure on F.

Using that $\hat{C}_\hbar Op_\hbar(\chi)$ are roughly orthogonal projectors with orthogonal images for distinct cylinders $\mathcal{C}$, this leads to:

LEMMA 11.19. *(see [**A**], (2.4.5)) Continue with the notation of Lemma 11.18. For $N = \kappa|\log\hbar|$, we have*

$$\begin{cases} \sum_{\mathcal{C}\in\Sigma_N} |\mu_\hbar(\mathcal{C})| = 1 + O(\hbar^\delta), \\ \\ \sum_{\mathcal{C}\in\Sigma_N(W_n,\tau)^c} |\mu_\hbar(\mathcal{C})| \geq \left(\frac{\kappa}{\vartheta}\theta\right)^2 + O(\hbar^\kappa), \\ \\ \implies \sum_{\mathcal{C}\in\Sigma_N(W_n,\tau)} |\mu_\hbar(\mathcal{C})| \leq 1 - \left(\frac{\kappa}{\vartheta}\theta\right)^2 + O(\hbar^\kappa). \end{cases}$$

The second inequality of Lemma 11.19 again goes in the opposite direction to the defining inequality (131) of a $(\hbar, 1-\theta, n)$ cover.

11.20.4. *Completion of Proof.* We now sketch the proof of the Theorem (see [**A**], (2.5.1)–(2.5.5)). It is similar to (124) but circumvents the positivity problem by decreasing N to below the Ehrenfest time.

PROOF. (Outline, quoted almost verbatim from [**A**])
As above, the quantum measure $\mu_\hbar$ is shift invariant (see [**A**], Proposition 1.3.1 (ii).)

We now fix n and let $N = \kappa|\log \hbar|$. The crucial point is that this is less than Ehrenfest time, unlike the choice of N in the first attempt. As above, $W_n \subset \Sigma_n$ denotes an $(\hbar, 1-\theta, n)$- cover of minimal cardinality of F by n-cylinders. In the following, we use the notation

$$\mu_\hbar(W_n) := \sum_{\mathcal{C}\in W_n} \mu_\hbar(\mathcal{C}), \quad \mu_\hbar(\Sigma_N(W_n, \tau)) := \sum_{\mathcal{C}\in\Sigma_N(W_n,\tau)} \mu_\hbar(\mathcal{C}). \tag{134}$$

We then run through the steps of (124) again;

$$\begin{aligned}
|\mu_\hbar(W_n)| &= |\frac{1}{N-n}\sum_{k=0}^{N-n-1} \mu_\hbar(\sigma^{-k}W_n)| \\
&= \left|\mu_\hbar\left(\frac{1}{N-n}\sum_{k=0}^{N-n-1} \mathbf{1}_{\sigma^{-k}W_n}\right)\right| \\
&\leq \sum_{\mathcal{C}\in\Sigma_N(W_n,\tau)} |\mu_\hbar(\mathcal{C})| + \tau \sum_{\mathcal{C}\notin\Sigma_N(W_n,\tau)} |\mu_\hbar(\mathcal{C})| \\
&\leq \sum_{C\in\Sigma_N(W_n,\tau)} |\mu_\hbar(\mathcal{C})| + \tau + O(h^\kappa) \\
&\leq (1-\tau)\left(1-(\frac{\kappa}{\vartheta}\theta)^2\right) + \tau + O(h^\kappa).
\end{aligned} \tag{135}$$

Above, Lemma 11.19 is used to circumvent the positivity problem in (124). We also used (again) (126) and the main estimate (123) to bound the N-cylinder sets in $\Sigma_N(W_n, \tau)$ with $N = \kappa|\log \hbar|$, but with the time n fixed.

Now let $\hbar \to 0$. Since W_n is fixed, the weak convergence $\mu_\hbar \to \mu_0$ implies,

$$\mu_0(W_n) \;\leq\; (1-\tau)\left(1-(\tfrac{\kappa}{\vartheta}\theta)^2\right) + \tau. \tag{136}$$

Since $F \subset W_n$, the same estimate applies to F. Since it holds for all $\theta < 1$, it follows that

$$\mu_0(F) \leq (1-\tau)\left(1-(\frac{\kappa}{\vartheta})^2\right) + \tau = 1 - (\frac{\kappa}{\vartheta})^2(1-\tau) < 1. \tag{137}$$

This completes the proof that supp $\nu_0 \neq F$. QED

To obtain the measure statement of Theorem 11.5, let

$$I_\delta = \{x \in S^*M : h_g(\mu_0^x) \leq \frac{\Lambda}{2}(1-\delta)\}. \tag{138}$$

It follows from the Shannon-McMillan theorem that for all $\alpha > 0$, there exists $I_\delta^\alpha \subset I_\delta$ with $\mu_0(I_\delta \backslash I_\delta^\alpha) \leq \alpha$ such that I_δ^α can be covered by $e^{n(\frac{\Lambda}{2}(1-\delta+\alpha)}$ n-cylinders for sufficiently large n. It follows from (137) that

$$\mu_0(I_\delta^\alpha) \leq (1-\tau(\delta-\alpha))\left(1-(\frac{\kappa}{\vartheta(\delta-\alpha)})_+^2\right) + \tau(\delta-\alpha). \tag{139}$$

Let $\alpha \to 0$ and one has,

$$\nu_0(S^*M\backslash I_\delta) \geq (1-\tau(\delta))(\frac{\kappa}{\vartheta(\delta)})^2. \tag{140}$$

□

11.21. $F = \gamma$**.** To help digest the proof, let us run through the proof in the simplest case where F is a single closed hyperbolic geodesic γ and M is a compact hyperbolic manifold of constant curvature -1. We would like to show that $\nu_0 \neq \mu_\gamma$, or equivalently that $\nu_0(\gamma) < 1$. What we need to show is that, if $\nu_0(F) = 1$, then it requires a lot of cylinder sets to cover F. Hence we try to get a contradiction from the fact that only a few cylinder sets are needed to cover γ. In fact, for each n, γ may be covered by one n-cylinder

$$W_n^0(\gamma) = \{[\alpha_0, \ldots, \alpha_0, \ldots, \alpha_n]\},$$

namely the one specified by the indices of the cells P_α that γ passes through starting from some fixed $x_0 \in \gamma$. To obtain a tube around the orbit, we take the union $W_n(\gamma) := \bigcup_{j=0}^{n-1} g^{-j} W_n^0(\gamma)$. We could use the two-sided symbols, going symmetrically forward and backward in time to obtain a transverse ball of radius $e^{-n/2}$.

First, let us try a naive heuristic argument. The main estimate (113) gives

$$|\mu_\hbar(W_n(\gamma))| \leq Ch^{-d/2}e^{-nd/2}(1+O(\epsilon))^n, \quad \text{uniformly for } n \leq K\,|\log \hbar|, \tag{141}$$

for any $K > 0$. It is now tempting to put $n = (1+\epsilon)|\log \hbar|$ with $\epsilon > 0$ to obtain $|\mu_\hbar(W_{n(\hbar,\theta)})| \leq \hbar^\theta$. But we cannot take the semi-classical limit of this estimate. If we did so formally (not rigorously), it would seem to imply that $\nu_0(\gamma) = 0$, which is stronger than the rigorous result $\nu_0(\gamma) < 1$ and possibly false in some examples (if weak scarring in fact occurs).

The problem is that the classical set $W_{n(\hbar,\theta)}$ is a kind of $\hbar$-dependent Bowen ball or tube around γ of radius $e^{-\frac{1}{2}n(\hbar,\theta)} = \hbar^{1/2+\theta}$. The uncertainty principle prohibits microlocalization to tubes of radius $< h^{\frac{1}{2}}$ and (as discussed in §5), it is difficult even to work with neighborhoods smaller than $h^{\frac{1}{2}-\delta}$ or at best $h^{\frac{1}{2}}|\log \hbar|$. In particular, we do not have control over the weak limits of $|\mu_\hbar(W_{n(\hbar,\theta)})|$, and there is no obvious reason why it should have semi-classical asymptotics. It is not a positive measure on this length scale, so we do not know if it implies $\nu_0(\gamma) = 0$. On the other hand, if we decrease n below the Ehrenfest time, so that the matrix element has a classical limit, then the bound is trivial.

The strategy of [**A**] is to use the sub-multiplicative estimate for $(\hbar, 1-\theta, N)$-covers of γ of minimal cardinality to decrease the length N below the Ehrenfest time (see Lemma 11.17). We first consider the cover $W_n := W_n(\gamma)$, which is a minimal cover of γ by cylinders of length n. We then form $\Sigma_N(W_n, \tau)$. At least intuitively, W_n corresponds to a Bowen ball of radius $e^{-n/2}$, i.e. a transverse cube of this radius in the stable and unstable directions. The cylinders of $\Sigma_N(W_n, \tau)$ satisfy $\sigma^j(\mathcal{C}) \cap W_n = \mathcal{C}$ for a proportion τ of $0 \leq j \leq N$. This carves out a very porous set in the stable/unstable cross section, suggesting why the lower bound in Lemma 11.19 should be true.

It would be interesting to connect the argument of Theorem 11.5 in this case to the problem in §5 (see §5.1) of estimating the eigenfunction mass inside a shrinking tubular neighborhood of γ, i.e. to give an upper bound for the mass $(Op_\hbar(\chi_1^\delta)\varphi_\mu, \varphi_\mu))$ in the shrinking tube around γ or (better) the mass $(Op_\hbar(\chi_2^\delta)\varphi_\mu, \varphi_\mu)$ in a shrinking phase space tube. However, the methods of [**A**] apply equally to the cat map setting and in that case, there are eigenfunctions whose mass in shrinking tubes around a hyperbolic fixed point have the same order of magnitude as in the integrable case up to a factor of 1/2 [**FNB**]. So unless one can use a property of Δ-eigenfunctions which is not shared by quantum cat map eigenfunctions, one cannot expect to prove a sharper upper bound for eigenfunction mass in shrinking tubes around hyperbolic orbits in the Anosov case than one has in the integrable case.

11.22. Some ideas on the main estimate. The main estimate is independently useful. For instance, it has been used to obtain bounds on decay rates on quantum scattering [**NZ**]. In this section, we sketch some ideas of the proof. We do not go so far as to explain a key point, that the estimate is valid for times $t \leq \kappa |\log \hbar|$ for any κ.

The goal is to estimate the norm of the operators $P^*_{\alpha'}\, U^n\, P_\alpha \operatorname{Op}(\chi^{(n)})$ in terms of $\hbar$ and n. Since $U(n)$ is unitary it suffices to estimate the norm of the cylinder set operators

$$K_n := \hat{P}_{\alpha_n} U \hat{P}_{\alpha_{n-1}} \cdots U \hat{P}_{\alpha_0} Op(\chi).$$

Let $K_n(z, w)$ be the Schwartz kernel of this operator. By the Schur inequality, the operator norm is bounded above by

$$\sup_z \int_M |K_n(z, w)| dV(w).$$

The inequality only applies when $K_n(z, w) \in L^1(M)$, but by harmless smoothing approximation we may assume this.

We may get an intuitive idea of the exponential decay of the norm of the cylinder set operators as follows. The wave group (or the semi-classical

Schrödinger propagator) U^t on a manifold without conjugate points can be lifted to the universal cover (see (25)) where it has a global Hadamard parametrix (17). This parametrix shows that the wave kernel on the universal cover is closely related to the spherical means operator $L_t f(x) = \int_{S_x^* M} f(g^t(x, \omega) d\mu_x(\omega)$, where $d\mu_x$ is the Euclidean surface measure on $S_x^* M$ induced by the metric. Both operators have the same wave front relation, and indeed the wave front set of $U(t, \cdot, y)$ is precisely the set of (co-)normal directions to the distance sphere $S_t(y)$ centered at y. The operator relating U^t and L_t is a dth order (pseudo-differential) operator in (t, x), and in the Fourier transformed picture in [**A**] it gives rise to the factor $h^{-d/2}$. As time evolves, the surface volume of the distance sphere in the universal cover grows exponentially. Since U^t is unitary, its amplitude must decay exponentially. In fact this is evident from (17) since $\Theta(x, y)$ grows exponentially in the distance from x to y, so $\Theta^{-1/2}(x, y)$ decays exponentially.

The cylinder set operators $\hat{\mathcal{C}}_\hbar = \hat{P}_{\alpha_n}(n) \cdots \hat{P}_{\alpha_1}(1) \hat{P}_{\alpha_0}(0)$ are given by a sequence of alternating applications of the propagator U and the cutting off operators P_{α_j} (which are multiplications). At least heuristically, in estimating the Schur integral for $K_n(x, y)$, one is first cutting off to a piece of the distance of the distance sphere $S_1(y)$ of radius one centered at y. Then one applies U and this piece expands to $S_2(y)$. Then one cuts it off with P_{α_2}. As this picture evolves, one ends up after n iterates with a piece of the distance sphere $S_n(y)$ of radius n centered at y which is has the same size as the initial piece. But the spherical means and wave group are normalized by dividing by the volume, so the contribution to the Schur integral is exponentially decaying at the rate e^{-n}.

11.23. Clarifications to [A].

(1) The θ on p. 447 is a new parameter < 1 which is implicitly defined above (2.0.1).
(2) The θ in (2.3.2) is also implicitly defined by this inequality. W_{n_1} is not assumed to be a θ cover.
(3) In (2.5.1)–(2.5.5), the Σ in $\sigma^{-k} \Sigma(W_{n_1})$ should be removed.
(4) As mentioned above, the notation $\mu_h(\Sigma_N(W_n, \tau))$ is not quite defined but means $\mu_h(\bigcup_{\mathcal{C} \in \Sigma_N(W_n, \tau))} \mathcal{C})$.

12. Applications to nodal hypersurfaces of eigenfunctions

This survey has been devoted to the asymptotics of matrix elements $\langle A\varphi_j, \varphi_j \rangle$. In the final section, we address the question, "what applications does the study of these matrix elements have?"

In automorphic forms, the matrix elements are related to L-functions, Rankin-Selberg zeta functions, etc. So there is ample motivation for their study in the arithmetic case. Such applications lie outside the scope of this survey, so we refer the reader to Sarnak's expository articles [**Sar2, Sar3**] for background on and references to the large literature in arithmetic quantum chaos.

Here we are concerned with applications to the classical and geometric analysis of eigenfunctions on general Riemannian manifolds. We present an application of quantum ergodicity to the study of nodal (zero) sets of eigenfunctions on real analytic (M, g) with ergodic geodesic flow. It is not implausible that one could use related methods to count critical points of eigenfunctions in the ergodic case.

The nodal hypersurface of φ_j is the zero set $\mathcal{N}_j = \{x \in M : \varphi_j(x) = 0\}$. We are interested in its distribution as $\lambda_j \to \infty$. We study the distribution through the integrals

$$\int_{\mathcal{N}_j} f dS$$

of f over the nodal line with respect to the natural surface measure. If $f = 1_U$ is the characteristic (indicator) function of an open set $U \subset M$, then

$$\int_{\mathcal{N}_j} 1_U ds = \mathcal{H}^{n-1}(U \cap \mathcal{N}_j) \tag{142}$$

is $(n-1)$ dimensional Hausdorff measure of the part of $\mathcal{N}_j$ which intersects U. When $U = M$ the integral (142) gives the total surface volume of the nodal set. The principal result on volumes is due to Donnelly-Fefferman [**DF**], proving a conjecture of S. T. Yau [**Y1**]:

Theorem 12.1. [**DF**] *(see also* [**Lin**]*) Let (M, g) be a compact real analytic Riemannian manifold, with or without boundary. Then there exist c_1, C_2 depending only on (M, g) such that*

$$c_1\lambda \leq \mathcal{H}^{m-1}(Z_{\varphi_\lambda}) \leq C_2\lambda, \quad (\Delta\varphi_\lambda = \lambda^2\varphi_\lambda; c_1, C_2 > 0). \tag{143}$$

The question is whether one can obtain asymptotics results by imposing a global dynamical condition on the geodesic flow.

The hypothesis of real analyticity is crucial: The analysis is based on the analytic continuation of eigenfunctions to the complexification $M_{\mathbb{C}}$ of M. The articles [**DF, Lin**] also used analytic continuation to study volumes of nodal hypersurfaces. It is the role of ergodicity in the complex setting that gives strong equidistribution results.

12.1. Complexification of (M, g) and Grauert tubes. By a theorem of Bruhat and Whitney, a real analytic manifold M always possesses a unique complexification $M_{\mathbb{C}}$, an open complex manifold in which M embeds $\iota : M \to M_{\mathbb{C}}$ as a totally real submanifold. Examples are real algebraic subvarieties of $\mathbb{R}^n$ which complexify to complex algebraic subvarieties of $\mathbb{C}^n$. Other simple examples include the complexification of a quotient such as $M = \mathbb{R}^m/\mathbb{Z}^m$ to $M_{\mathbb{C}} = \mathbb{C}^m/\mathbb{Z}^m$. The complexification of hyperbolic space can be defined using the hyperboloid model $\mathbb{H}^n = \{x_1^2 + \cdots x_n^2 - x_{n+1}^2 = -1, \ x_n > 0\}$. Then, $\mathbb{H}^n_{\mathbb{C}} = \{(z_1, \ldots, z_{n+1}) \in \mathbb{C}^{n+1} : z_1^2 + \cdots z_n^2 - z_{n+1}^2 = -1\}$.

The Riemannian metric determines a special kind of plurisubharmonic exhaustion function on $M_{\mathbb{C}}$ [**GS1, GS2, LemS1**], namely the Grauert tube radius $\sqrt{\rho} = \sqrt{\rho}_g$ on $M_{\mathbb{C}}$, defined as the unique solution of the Monge-Ampère equation

$$(\partial\bar{\partial}\sqrt{\rho})^m = \delta_{M_{\mathbb{R}},dV_g}, \quad \iota^*(i\partial\bar{\partial}\rho) = g.$$

Here, $\delta_{M_{\mathbb{R}},dV_g}$ is the delta-function on the real M with respect to the volume form dV_g, i.e. $f \to \int_M f dV_g$. It is shown in [**GS1**] that $\sqrt{\rho}(\zeta) = i\sqrt{r^2(\zeta, \bar{\zeta})}$ where $r^2(x, y)$ is the squared distance function in a neighborhood of the diagonal in $M \times M$.

The Grauert tubes M_τ is defined by

$$M_\tau = \{\zeta \in M_{\mathbb{C}} : \sqrt{\rho}(\zeta) \leq \tau\}.$$

The maximal value of τ_0 for which $\sqrt{\rho}$ is well defined is known as the Grauert tube radius. For $\tau \leq \tau_0$, M_τ is a strictly pseudo-convex domain in $M_{\mathbb{C}}$. Using the complexified exponential map $(x, \xi) \to exp_x i\xi$ one may identify M_τ with the co-ball bundle $B^*_\tau M$ to M_τ. Under this identification, $\sqrt{\rho}$ corresponds to $|\xi|_g$. The one-complex dimensional null foliation of $\partial\bar{\partial}\sqrt{\rho}$, known as the "Monge-Ampère" or Riemann foliation, are the complex curves $t + i\tau \to \tau\dot{\gamma}(t)$, where γ is a geodesic. The geometric Grauert tube radius is the maximal radius for which the exponential map has a holomorphic extension and defines a diffeomorphism.

12.2. Analytic Continuation of eigenfunctions. The eigenfunctions are real analytic and therefore possess analytic continuations to some Grauert tube M_ϵ independent of the eigenvalue. To study analytic continuation of eigenfunctions, it is useful to relate them to analytic continuation of the wave group at imaginary time, $U(i\tau, x, y) = e^{-\tau\sqrt{\Delta}}(x, y)$. This is the Poisson operator of (M, g). We denote its analytic continuation in the x variable to M_ϵ by $U_{\mathbb{C}}(i\tau, \zeta, y)$. In terms of the eigenfunction expansion, one has

$$U(i\tau, \zeta, y) = \sum_{j=0}^{\infty} e^{-\tau\lambda_j} \varphi_j^{\mathbb{C}}(\zeta)\varphi_j(y), \quad (\zeta, y) \in M_\epsilon \times M. \tag{144}$$

To the author's knowledge, the largest ϵ has not been determined at this time. In particular, it is not known if the radius of analytic continuation of the wave kernel and of the eigenfunctions is the same as the geometrically defined Grauert tube radius.

Since

$$U_{\mathbb{C}}(i\tau)\varphi_\lambda = e^{-\tau\lambda}\varphi_\lambda^{\mathbb{C}}, \tag{145}$$

the analytic continuability of the Poisson operator to M_τ implies that every eigenfunction analytically continues to the same Grauert tube. It follows that the analytic continuation operator to M_τ is given by

$$A_{\mathbb{C}}(\tau) = U_{\mathbb{C}}(i\tau) \circ e^{\tau\sqrt{\Delta}}. \tag{146}$$

Thus, a function $f \in C^\infty(M)$ has a holomorphic extension to the closed tube $\sqrt{\rho}(\zeta) \leq \tau$ if and only if $f \in Dom(e^{\tau\sqrt{\Delta}})$, where $e^{\tau\sqrt{\Delta}}$ is the backwards "heat operator" generated by $\sqrt{\Delta}$ (rather than Δ).

Let us consider examples of holomorphic continuations of eigenfunctions:

- On the flat torus $\mathbb{R}^m/\mathbb{Z}^m$, the real eigenfunctions are $\cos\langle k, x\rangle$, $\sin\langle k, x\rangle$ with $k \in 2\pi\mathbb{Z}^m$. The complexified torus is $\mathbb{C}^m/\mathbb{Z}^m$ and the complexified eigenfunctions are $\cos\langle k, \zeta\rangle, \sin\langle k, \zeta\rangle$ with $\zeta = x + i\xi$.
- On the unit sphere S^m, eigenfunctions are restrictions of homogeneous harmonic functions on $\mathbb{R}^{m+1}$. The latter extend holomorphically to holomorphic harmonic polynomials on $\mathbb{C}^{m+1}$ and restrict to holomorphic function on $S^m_{\mathbb{C}}$.
- On $\mathbf{H}^m$, one may use the hyperbolic plane waves $e^{(i\lambda+1)\langle z,b\rangle}$, where $\langle z, b\rangle$ is the (signed) hyperbolic distance of the horocycle passing through z and b to 0. They may be holomorphically extended to the maximal tube of radius $\pi/4$. See e.g. [**BR**].
- On compact hyperbolic quotients $\mathbf{H}^m/\Gamma$, eigenfunctions can be then represented by Helgason's generalized Poisson integral formula,

$$\varphi_\lambda(z) = \int_B e^{(i\lambda+1)\langle z,b\rangle} dT_\lambda(b).$$

  To analytically continue φ_λ it suffices to analytically continue $\langle z, b\rangle$. Writing the latter as $\langle \zeta, b\rangle$, we have:

$$\varphi_\lambda^{\mathbb{C}}(\zeta) = \int_B e^{(i\lambda+1)\langle \zeta,b\rangle} dT_\lambda(b). \tag{147}$$

- In the case of 1D Schrödinger operators, several authors have studied analytic continuations of eigenfunctions and their complex zeros, see e.g. [**Hez, EGS**].

12.3. Maximal plurisubharmonic functions and growth of $\varphi_\lambda^{\mathbb{C}}$. In the case of domains $\Omega \subset \mathbb{C}^m$, the maximal PSH (pluri-subharmonic) function (or pluri-complex Green's function) relative to a subset $E \subset \Omega$ is defined by

$$V_E(\zeta) = \sup\{u(z) : u \in PSH(\Omega), u|_E \leq 0, u|_{\partial\Omega} \leq 1\}.$$

An alternative definition due to Siciak takes the supremum only with respect to polynomials p. We denote by $\mathcal{P}^N$ the space of all complex analytic polynomials of degree N and put $\mathcal{P}^N_K = \{p \in \mathcal{P}^N : ||p||_K \leq 1, \ ||p||_\Omega \leq e\}$. Then define

$$\log \Phi^N_E(\zeta) = \sup\{\frac{1}{N}\log|p_N(\zeta)| : p \in \mathcal{P}^N_E\}, \quad \log\Phi_E = \limsup_{N\to\infty} \log\Phi^N_E.$$

Here, $||f||_K = \sup_{z \in K} |f(z)|$. Siciak proved that $\log \Phi_E = V_E$.

On a real analytic Riemannian manifold, the natural analogue of $\mathcal{P}^N$ is the space

$$\mathcal{H}^\lambda = \{p = \sum_{j:\lambda_j \leq \lambda} a_j \varphi_{\lambda_j}, \ \ a_1, \dots, a_{N(\lambda)} \in \mathbb{R}\}$$

spanned by eigenfunctions with frequencies $\leq \lambda$. Rather than using the sup norm, it is convenient to work with L^2 based norms than sup norms, and so we define

$$\mathcal{H}^\lambda_M = \{p = \sum_{j:\lambda_j \leq \lambda} a_j \varphi_{\lambda_j}, \ \ ||p||_{L^2(M)} = \sum_{j=1}^{N(\lambda)} |a_j|^2 = 1\}.$$

We define the λ-Siciak approximate extremal function by

$$\Phi^\lambda_M(z) = \sup\{|\psi(z)|^{1/\lambda} \colon \psi \in \mathcal{H}_\lambda; \|\psi\|_M \leqslant 1\},$$

and the extremal function by

$$\Phi_M(z) = \sup_\lambda \Phi^\lambda_M(z).$$

It is not hard to show that

$$\Phi_M(z) = \lim_{\lambda \to \infty} \frac{1}{\lambda} \log \Pi_{[0,\lambda}(\zeta, \bar{\zeta}) = \sqrt{\rho}. \tag{148}$$

12.4. Nodal hypersurfaces in the case of ergodic geodesic flow. The complex nodal hypersurface of an eigenfunction is defined by

$$Z_{\varphi_\lambda^{\mathbb{C}}} = \{\zeta \in B^*_{\epsilon_0} M : \varphi_\lambda^{\mathbb{C}}(\zeta) = 0\}. \tag{149}$$

There exists a natural current of integration over the nodal hypersurface in any ball bundle $B^*_\epsilon M$ with $\epsilon < \epsilon_0$, given by

(150)

$$\langle [Z_{\varphi_\lambda^{\mathbb{C}}}], \varphi \rangle = \frac{i}{2\pi} \int_{B^*_\epsilon M} \partial\bar{\partial} \log |\varphi_\lambda^{\mathbb{C}}|^2 \wedge \varphi = \int_{Z_{\varphi_\lambda^{\mathbb{C}}}} \varphi, \quad \varphi \in \mathcal{D}^{(m-1,m-1)}(B^*_\epsilon M).$$

In the second equality we used the Poincaré-Lelong formula. The notation $\mathcal{D}^{(m-1,m-1)}(B^*_\epsilon M)$ stands for smooth test $(m-1, m-1)$-forms with support in $B^*_\epsilon M$.

The nodal hypersurface $Z_{\varphi_\lambda^{\mathbb{C}}}$ also carries a natural volume form $|Z_{\varphi_\lambda^{\mathbb{C}}}|$ as a complex hypersurface in a Kähler manifold. By Wirtinger's formula, it equals the restriction of $\frac{\omega_g^{m-1}}{(m-1)!}$ to $Z_{\varphi_\lambda^{\mathbb{C}}}$.

THEOREM 12.2. *[Z5] Let (M, g) be real analytic, and let $\{\varphi_{j_k}\}$ denote a quantum ergodic sequence of eigenfunctions of its Laplacian Δ. Let $(B^*_{\epsilon_0} M, J)$ be the maximal Grauert tube around M with complex structure J_g adapted to g. Let $\epsilon < \epsilon_0$. Then:*

$$\frac{1}{\lambda_{j_k}} [Z_{\varphi_{j_k}^{\mathbb{C}}}] \to \frac{i}{\pi} \partial\bar{\partial} \sqrt{\rho} \ \ \textit{weakly in} \ \ \mathcal{D}'^{(1,1)}(B^*_\epsilon M),$$

in the sense that, for any continuous test form $\psi \in \mathcal{D}^{(m-1,m-1)}(B_\epsilon^* M)$, *we have*

$$\frac{1}{\lambda_{j_k}} \int_{Z_{\varphi_{j_k}^{\mathbb{C}}}} \psi \to \frac{i}{\pi} \int_{B_\epsilon^* M} \psi \wedge \partial\bar{\partial}\sqrt{\rho}.$$

Equivalently, for any $\varphi \in C(B_\epsilon^* M)$,

$$\frac{1}{\lambda_{j_k}} \int_{Z_{\varphi_{j_k}^{\mathbb{C}}}} \varphi \frac{\omega_g^{m-1}}{(m-1)!} \to \frac{i}{\pi} \int_{B_\epsilon^* M} \varphi \partial\bar{\partial}\sqrt{\rho} \wedge \frac{\omega_g^{m-1}}{(m-1)!}.$$

COROLLARY 12.3. *Let* (M, g) *be a real analytic with ergodic geodesic flow. Let* $\{\varphi_{j_k}\}$ *denote a full density ergodic sequence. Then for all* $\epsilon < \epsilon_0$,

$$\frac{1}{\lambda_{j_k}} [Z_{\varphi_{j_k}^{\mathbb{C}}}] \to \frac{i}{\pi} \partial\bar{\partial}\sqrt{\rho}, \quad \textit{weakly in } \mathcal{D}'^{(1,1)}(B_\epsilon^* M).$$

REMARK: The limit form is singular with respect to the Kähler form $i\partial\bar{\partial}\rho$ on $M_{\mathbb{C}}$ adapted to the metric. Its highest exterior power is the delta function on the zero section (i.e. the real domain M).

In outline the steps in the proof are;

(1) By the Poincaré-Lelong formula, $[Z_{\varphi_\lambda^{\mathbb{C}}}] = i\partial\bar{\partial} \log |\varphi_\lambda^{\mathbb{C}}|$. This reduces the theorem to determining the limit of $\frac{1}{\lambda} \log |\varphi_\lambda^{\mathbb{C}}|$.
(2) $|\varphi_\lambda^{\mathbb{C}}|^2$, when properly L^2 normalized on each ∂M_τ is a quantum ergodic sequence on ∂M_τ.
(3) The ergodic property of complexified eigenfunctions implies that the L^2 norm of $|\varphi_\lambda^{\mathbb{C}}|^2$ on ∂M_τ is asymtotically $\sqrt{\rho}$. Thus, normalized log moduli of ergodic eigenfunctions are asympotically maximal PSH functions. They are λ-Siciak extremal functions.
(4) $\frac{1}{\lambda} \log \frac{|\varphi_\lambda^{\mathbb{C}}(\zeta)|^2}{||\varphi_\lambda^{\mathbb{C}}||^2_{L^2(\partial M_\tau}} \to 0$ for an ergodic sequence of eigenfunctions. Hence,
(5) $\frac{i}{\lambda} \partial\bar{\partial} \log |\varphi_\lambda^{\mathbb{C}}| \to i\partial\bar{\partial}\sqrt{\rho}$,

concluding the proof.

REMARK: We expect a similar result for nodal lines of Dirichlet or Neumann eigenfunctions of a piecewise smooth domain $\Omega \subset \mathbb{R}^2$ with ergodic billiards. We also expect similar results for nodal lines of such domains which touch the boundary (like the angular "spokes" of eigenfunctions of the disc). In [**TZ1**], we proved that the number of such nodal lines is bounded above by $C_\Omega \lambda$. This is rather like the Donnelly-Fefferman upper bound. However the example of the disc shows that there is no lower bound above zero. That is, there are sequences of eigenfunctions (with low angular momentum) with a uniformly bounded number of nodal lines touching the boundary. We believe that ergodicity is a mechanism producing lower bounds on numbers of zeros and critical points. It causes eigenfunctions to oscillate uniformly in all directions and hence to produce a lot of zeros and critical points. We expect this to be one of the future roles for ergodicity of eigenfunctions.

REMARK: Precursors to Theorem 12.2 giving limit distributions of zero sets for holomorphic eigen-sections of Kähler quantizations of ergodic symplectic maps on Kähler manifolds can be found in [**NV2, SZ**]. In that setting, the zeros become uniformly distributed with respect to the Kähler form. In the Riemannian setting, they become uniformly distributed with respect to a singular $(1,1)$ current relative to the Kähler form $\partial\bar{\partial}\rho$ on $M_{\mathbb{C}}$.

References

[A] N. Anantharaman, Entropy and the localization of eigenfunctions, Ann. of Math. (2) 168 (2008), no. 2, 435–475.

[A2] N. Anantharaman, Eigenfunctions of the Laplacian on negatively curved manifolds: a semi-classical approach, Lecture Notes (online at http://www.math.polytechnique.fr/ nalini/).

[ANK] N. Anantharaman, H. Koch, and S. Nonnenmacher, Entropy of eigenfunctions, arXiv:0704.1564.

[AN] N. Anantharaman and S. Nonnenmacher, Half-delocalization of eigenfunctions for the Laplacian on an Anosov manifold, Annales Inst. Fourier 57, number 6 (2007), 2465-2523 (arXiv:math-ph/0610019).

[AN2] N. Anantharaman and S. Nonnenmacher, Vibrations chaotiques et grosses balafres. Gaz. Math. No. 119 (2009), 16–32. English Version (different): Chaotic vibrations and strong scars.

[AZ] N. Anantharaman and S. Zelditch, Patterson-Sullivan distributions and quantum ergodicity. Ann. Henri Poincar 8 (2007), no. 2, 361–426.

[AS] R. Aurich and F. Steiner, Statistical properties of highly excited quantum eigenstates of a strongly chaotic system. Phys. D 64 (1993), no. 1-3, 185–214.

[B.B] V. M. Babic and V. S. Buldyrev, *Short-Wavelength Diffraction Theory*, Springer Series on Wave Phenomena 4, Springer-Verlag, New York (1991).

[BSS] A. Backer, R. Schubert, and P. Stifter, On the number of bouncing ball modes in billiards. J. Phys. A 30 (1997), no. 19, 6783–6795.

[Bar] A. Barnett, Asymptotic rate of quantum ergodicity in chaotic Euclidean billiards, Comm. Pure Appl. Math. 59 (2006), no. 10, 1457–1488.

[Be] P. Bérard, On the wave equation without conjugate points, Math. Zeit. 155 (1977), 249–276.

[BR] J. Bernstein and A. Reznikov, Analytic continuation of representations and estimates of automorphic forms. Ann. of Math. (2) 150 (1999), no. 1, 329–352.

[Ber] M. Berry, Regular and irregular semiclassical wavefunctions, J. Phys. A 10 (1977), 2083–2091.

[BL] J. Bourgain and E. Lindenstrauss, Entropy of quantum limits. Comm. Math. Phys. 233 (2003), no. 1, 153–171.

[BouR] A. Bouzouina and D. Robert, Uniform semiclassical estimates for the propagation of quantum observables. Duke Math. J. 111 (2002), no. 2, 223–252.

[Bu] N. Burq, Quantum ergodicity of boundary values of eigenfunctions: A control theory approach, Canad. Math. Bull. 48 (2005), no. 1, 3–15 (math.AP/0301349).

[BZ] N. Burq and M. Zworski, Geometric control in the presence of a black box. J. Amer. Math. Soc. 17 (2004), no. 2, 443–471.

[Chr] H. Christianson, Semiclassical non-concentration near hyperbolic orbits. J. Funct. Anal. 246 (2007), no. 2, 145–195.

[CV] Y. Colin de Verdière, Ergodicité et fonctions propres du Laplacien, Comm. Math. Phys. 102 (1985), 497-502.

[CV2] Y. Colin de Verdière, Quasi-modes sur les varietes Riemanniennes compactes, Invent. Math. 43 (1977), 15-52.

[CV3] Y. Colin de Verdière, Semi-classical measures and entropy [after Nalini Anantharaman and Stéphane Nonnenmacher]. Séminaire Bourbaki. Vol. 2006/2007. Astérisque No. 317 (2008), Exp. No. 978, ix, 393–414.

[CVP] Y. Colin de Verdière and B. Parisse, Équilibre instable en régime semi-classique. I. Concentration microlocale, Comm. in PDE 19 (1994), no. 9-10, 1535–1563.

[CR] M. Combescure and D. Robert, Semiclassical spreading of quantum wave packets and applications near unstable fixed points of the classical flow, Asymptot. Anal. 14 (1997), 377–404.

[DSj] M. Dimassi and J. Sjöstrand, *Spectral asymptotics in the semi-classical limit.* London Mathematical Society Lecture Note Series, 268. Cambridge University Press, Cambridge (1999).

[DF] H. Donnelly and C. Fefferman, Nodal sets of eigenfunctions on Riemannian manifolds, Invent. Math. 93 (1988), 161-183.

[Dui] J. J. Duistermaat, Oscillatory integrals, Lagrange immersions and unfolding of singularities. Comm. Pure Appl. Math. 27 (1974), 207–281.

[D.G] J.J.Duistermaat and V.Guillemin, The spectrum of positive elliptic operators and periodic bicharacteristics, Inv.Math. 24 (1975), 39-80.

[EGS] A. Eremenko, A. Gabrielov, and B. Shapiro, High energy eigenfunctions of one-dimensional Schrödinger operators with polynomial potentials. Comput. Methods Funct. Theory 8 (2008), no. 1-2, 513–529.

[EZ] L. Evans and M. Zworski, Introduction to semiclassical analysis, UC Berkeley lecture notes (2006). http://math.berkeley.edu/ zworski.

[FN] F. Faure and S. Nonnenmacher, On the maximal scarring for quantum cat map eigenstates. Comm. Math. Phys. 245 (2004), no. 1, 201–214.

[FNB] F. Faure, S. Nonnenmacher, and S. De Bièvre, Scarred eigenstates for quantum cat maps of minimal periods. Comm. Math. Phys. 239 (2003), no. 3, 449–492.

[FP] M. Feingold and A. Peres, Distribution of matrix elements of chaotic systems. Phys. Rev. A (3) 34 (1986), no. 1, 591–595.

[GL] P. Gérard and E. Leichtnam, Ergodic properties of eigenfunctions for the Dirichlet problem, Duke Math J. 71 (1993), 559-607.

[GSj] A. Grigis and J. Sjöstrand, *Microlocal analysis for differential operators*, London Math. Soc. Lecture Notes 196 (1994).

[GS1] V. Guillemin and M. Stenzel, Grauert tubes and the homogeneous Monge-Ampère equation. J. Differential Geom. 34 (1991), no. 2, 561–570.

[GS2] V. Guillemin and M. Stenzel, Grauert tubes and the homogeneous Monge-Ampère equation. II. J. Differential Geom. 35 (1992), no. 3, 627–641.

[GSt] V. Guillemin and S. Sternberg, *Geometric asymptotics.* Mathematical Surveys, No. 14. American Mathematical Society, Providence, R.I., (1977).

[Gut] B. Gutkin, Note on converse quantum ergodicity. Proc. Amer. Math. Soc. 137 (2009), no. 8, 2795–2800.

[HanL] Q. Han and F.H. Lin *Nodal sets of solutions of elliptic differential equations*, book in preparation (2007).

[Has] A. Hassell, Ergodic billiards that are not quantum unique ergodic, Appendix by A. Hassell and L. Hillairet, to appear in Ann. Math. (arXiv:0807.0666).

[HZ] A. Hassell and S. Zelditch, Quantum ergodicity of boundary values of eigenfunctions. Comm. Math. Phys. 248 (2004), no. 1, 119–168.

[HR] D. A. Hejhal and B. N. Rackner, On the topography of Maass waveforms for PSL$(2, Z)$. Experiment. Math. 1 (1992), no. 4, 275–305.

[Hel] S. Helgason, *Topics in harmonic analysis on homogeneous spaces.* Progress in Mathematics, 13. Birkhuser, Boston, Mass., (1981).

[He] E. J. Heller, Bound-state eigenfunctions of classically chaotic Hamiltonian systems: scars of periodic orbits. Phys. Rev. Lett. 53 (1984), no. 16, 1515–1518.

[He2] E. J. Heller, Wavepacket dynamics and quantum chaology. Chaos et physique quantique (Les Houches, 1989), 547–664, North-Holland, Amsterdam, (1991).

[HO] E. J. Heller and P. W. O'Connor, Quantum localization for a strongly classical chaotic system, Phys. Rev. Lett. 61 (20) (1988), 2288-2291.

[Hez] H. Hezari, Complex zeros of eigenfunctions of 1D Schrdinger operators. Int. Math. Res. Not. IMRN 2008, no. 3, Art. ID rnm148, 23 pp.

[Hol] R. Holowinsky, Sieving for mass equidistribution, to appear in Annals of Math. (arXiv:0809.1640).

[HS] R. Holowinsky and K. Soundararajan, Mass equidistribution for Hecke eigenforms, to appear in Annals of Math. (arXiv:0809.1636).

[HoI-IV] L. Hörmander, *Theory of Linear Partial Differential Operators I-IV*, Springer-Verlag, New York (1985).

[J] O. Jenkinson, Every ergodic measure is uniquely maximizing, Disc. Cont. Dyn. Syst. 16 (2006).

[Kl] W. Klingenberg, *Lectures on Closed Geodesics*, Grundlehren der. math. W. 230, Springer-Verlag (1978).

[KR] P. Kurlberg and Z. Rudnick, On the distribution of matrix elements for the quantum cat map. Ann. of Math. (2) 161 (2005), no. 1, 489–507.

[L] V. F. Lazutkin, *KAM theory and semiclassical approximations to eigenfunctions.* With an addendum by A. I. Shnirelman. Ergebnisse der Mathematik und ihrer Grenzgebiete (3), 24. Springer-Verlag, Berlin, (1993).

[LemS1] L. Lempert and R. Szöke, Global solutions of the homogeneous complex Monge-Ampère equation and complex structures on the tangent bundle of Riemannian manifolds. Math. Ann. 290 (1991), no. 4, 689–712.

[Lin] F.H. Lin, Nodal sets of solutions of elliptic and parabolic equations. Comm. Pure Appl. Math. 44 (1991), no. 3, 287–308.

[LIND] E. Lindenstrauss, Invariant measures and arithmetic quantum ergodicity, Annals of Math. (2) 163 (2006), no. 1, 165–219.

[LIND2] E. Lindenstrauss, Adelic dynamics and arithmetic quantum unique ergodicity. Current developments in mathematics, 2004, 111–139, Int. Press, Somerville, MA, (2006).

[LS] W. Luo and P. Sarnak, Quantum ergodicity of eigenfunctions on $\mathrm{PSL}_2(\mathbb{Z})\backslash\mathbf{H}^2$, IHES Publ. 81 (1995), 207-237.

[LS2] W. Luo and P. Sarnak, Quantum variance for Hecke eigenforms, Annales Scient. de l'École Norm. Sup. 37 (2004), p. 769-799.

[M] J. Marklof, Arithmetic quantum chaos, Encyclopedia of Mathematical Physics article 449.

[MOZ] J. Marklof and S. O'Keefe, Weyl's law and quantum ergodicity for maps with divided phase space. With an appendix "Converse quantum ergodicity" by Steve Zelditch. Nonlinearity 18 (2005), no. 1, 277–304.

[MS] S. D. Miller and W. Schmid, The highly oscillatory behavior of automorphic distributions for SL(2). Lett. Math. Phys. 69 (2004), 265–286.

[NJT] N. Nadirashvili, D. Jakobson, and J.A. Toth, Geometric properties of eigenfunctions. (Russian) Uspekhi Mat. Nauk 56 (2001), no. 6(342), 67–88; translation in Russian Math. Surveys 56 (2001), no. 6, 1085–1105.

[NV] S. Nonnenmacher and A. Voros, Eigenstate structures around a hyperbolic point. J. Phys. A 30 (1997), no. 1, 295–315.

[NV2] S. Nonnenmacher and A. Voros, Chaotic eigenfunctions in phase space. J. Statist. Phys. 92 (1998), no. 3-4, 431–518.

[NZ] N. Nonnenmacher and M. Zworski, Quantum decay rates in chaotic scattering, to appear in Acta. Math. (arXiv:0706.3242).

[O] J. P. Otal, Sur les fonctions propres du laplacien du disque hyperbolique. (French. English, French summary) [About eigenfunctions of the Laplacian on the hyperbolic disc] C. R. Acad. Sci. Paris Sér. I Math. 327 (1998), no. 2, 161–166.

[Ra] J. Ralston, Gaussian beams and the propagation of singularities. *Studies in partial differential equations,* 206–248, MAA Stud. Math., 23, Math. Assoc. America, Washington, DC, (1982).

[Ra2] J. Ralston, Approximate eigenfunctions of the Laplacian. J. Differential Geometry 12 (1977), no. 1, 87–100.

[Riv] G. Riviere, Entropy of semiclassical measures in dimension 2 (arXiv: 0809.0230).

[RS] Z. Rudnick and P. Sarnak, The behaviour of eigenstates of arithmetic hyperbolic manifolds. Comm. Math. Phys. 161 (1994), no. 1, 195–213.

[Ru] D. Ruelle, *Statistical mechanics: Rigorous results.* W. A. Benjamin, Inc., New York-Amsterdam (1969).

[Sar] P. Sarnak, Integrals of products of eigenfunctions. Internat. Math. Res. Notices no. 6 (1997), 251–261.

[Sar2] P. Sarnak, Arithmetic quantum chaos. The Schur lectures (1992) (Tel Aviv), 183–236, Israel Math. Conf. Proc., 8, Bar-Ilan Univ., Ramat Gan, (1995). 1926.

[Sar3] P. Sarnak, Recent progress on QUE, BAMS (to appear).

[Sch] W. Schmid, Automorphic distributions for $SL(2,\mathbb{R})$. (English summary) Confrence Mosh Flato 1999, Vol. I (Dijon), 345–387, Math. Phys. Stud., 21, Kluwer Acad. Publ., Dordrecht, (2000).

[Schu] R. Schubert, On the rate of quantum ergodicity for quantised maps. Ann. Henri Poincaré 9 (2008), no. 8, 1455–1477.

[Schu2] R. Schubert, Upper bounds on the rate of quantum ergodicity. Ann. Henri Poincaré 7 (2006), no. 6, 1085–1098.

[Schu3] R. Schubert, Semiclassical behaviour of expectation values in time evolved Lagrangian states for large times. Comm. Math. Phys. 256 (2005), no. 1, 239–254.

[Sh.1] A. I. Shnirelman, Ergodic properties of eigenfunctions, Usp. Math. Nauk. 29/6 (1974), 181-182.

[Sh.2] A. I. Shnirelman, On the asymptotic properties of eigenfunctions in the region of chaotic motion, addendum to V.F.Lazutkin, *KAM theory and semiclassical approximations to eigenfunctions*, Springer (1993).

[SZ] B. Shiffman and S. Zelditch, Distribution of zeros of random and quantum chaotic sections of positive line bundles. (English summary) Comm. Math. Phys. 200 (1999), no. 3, 661–683.

[Sound1] K. Soundararajan, Quantum unique ergodicity for $SL_2(Z)\mathbf{H}$, to appear in Annals of Math. (arXiv:0901.4060)

[Su] T. Sunada, Quantum ergodicity. Progress in inverse spectral geometry, 175–196, Trends Math., Birkhuser, Basel, 1997.

[TZ1] J. A. Toth and S. Zelditch, Counting nodal lines which touch the boundary of an analytic domain. J. Differential Geom. 81 (2009), no. 3, 649–686.

[TZ2] J. A. Toth and S. Zelditch, L^p norms of eigenfunctions in the completely integrable case. Ann. Henri Poincaré 4 (2003), no. 2, 343–368.

[TZ3] J. A. Toth and S. Zelditch, Quantum ergodic restriction theorems, I: interior hypersurfaces in domains with ergodic billiards (preprint, 2010).

[Wat] T. Watson, *Central Value of Rankin Triple L-function for Unramified Maass Cusp Forms,* to appear in Annals of Math.

[W] S. A. Wolpert, Semiclassical limits for the hyperbolic plane. Duke Math. J. 108 (2001), no. 3, 449–509.

[W2] S. A. Wolpert, Asymptotic relations among Fourier coefficients of automorphic eigenfunctions. Trans. Amer. Math. Soc. 356 (2004), no. 2, 427–456.

[W3] S. A. Wolpert, The modulus of continuity for $\Gamma_0(m)\backslash\mathbb{H}$ semi-classical limits, Comm. Math. Phys. 216 (2001), 313–323.

[Y1] S. T. Yau, Survey on partial differential equations in differential geometry. *Seminar on Differential Geometry*, pp. 3–71, Ann. of Math. Stud., 102, Princeton Univ. Press, Princeton, N.J., (1982).

[Z1] S. Zelditch, Local and global analysis of eigenfunctions, in: *Handbook of Geometric Analysis, No. 1* L. Ji, P. Li, R. Schoen and L. Simon (eds.), Somerville, MA: International Press; Beijing: Higher Education Press. Advanced Lectures in Mathematics (ALM) 7, 545–658 (2008).

[Z2] S. Zelditch, Quantum ergodicity and mixing, *Encyclopedia of Mathematical Physics* Vol. 4, Ed. J.P. Franoise, G. Naber, S.T. Tsou (2007), 183–196.

[Z3] S. Zelditch, Uniform distribution of eigenfunctions on compact hyperbolic surfaces. Duke Math. J. 55 (1987), no. 4, 919–941.

[Z4] S. Zelditch, On the rate of quantum ergodicity. I. Upper bounds. Comm. Math. Phys. 160 (1994), no. 1, 81–92.

[Z5] S. Zelditch, Complex zeros of real ergodic eigenfunctions. Invent. Math. 167 (2007), no. 2, 419–443.

[Z6] S. Zelditch, Quantum ergodicity of C^* dynamical systems. Comm. Math. Phys. 177 (1996), no. 2, 507–528.

[Z7] S. Zelditch, Note on quantum unique ergodicity, Proc. Amer. Math. Soc. 132 (2004), no. 6, 1869–1872.

[Z8] S. Zelditch, Mean Lindelöf hypothesis and equidistribution of cusp forms and Eisenstein series. J. Funct. Anal. 97 (1991), no. 1, 1–49.

[Z9] S. Zelditch, Quantum transition amplitudes for ergodic and for completely integrable systems. J. Funct. Anal. 94 (1990), no. 2, 415–436.

[Z10] S. Zelditch, On a "quantum chaos" theorem of R. Schrader and M. Taylor. J. Funct. Anal. 109 (1992), no. 1, 1–21.

[ZZw] S.Zelditch and M.Zworski, Ergodicity of eigenfunctions for ergodic billiards, Comm.Math. Phys. 175 (1996), 673-682.

[Zh] P. Zhao, Quantum Variance of Maass-Hecke Cusp Forms, to appear in Comm. Math. Phys.

Department of Mathematics, Johns Hopkins University, Baltimore, MD 21218, USA

E-mail address: zel@math.jhu.edu

Made in United States
Orlando, FL
04 November 2026

90200852R00116